Insects in Camera

Insects in Camera

A Photographic Essay on Behaviour

TEXT
Christopher O'Toole
Hope Entomological Collections, University Museum, Oxford

PHOTOGRAPHY
Ken Preston-Mafham
Premaphotos Wildlife, Alcester

OXFORD   NEW YORK   TORONTO
Oxford University Press · 1985

Oxford University Press, Walton Street, Oxford OX2 6DP

London New York Toronto
Delhi Bombay Calcutta Madras Karachi
Kuala Lumpur Singapore Hong Kong Tokyo
Nairobi Dar es Salaam Cape Town
Melbourne Auckland

and associated companies in
Beirut Berlin Ibadan Mexico City Nicosia

Oxford is a trade mark of Oxford University Press

British Library Cataloguing in Publication Data
O'Toole, Christopher
 Insects in camera.
 1. Insects—Behavior—Pictorial works
 I. Title II. Preston-Mafham, Ken
 595.7051'022'2 QL496
 ISBN 0–19–217694–3

Library of Congress Cataloging in Publication Data
O'Toole, Christopher.
 Insects in camera.
 Bibliography: p.
 Includes index.
 1. Insects—Behavior. I. Preston-Mafham, Ken.
 II. Title.
 QL496.086 1983 595.7'051 84–5648
 ISBN 0–19–217694–3

Set by Wyvern Typesetting Ltd, Bristol
Printed in Hong Kong

There would seem to be only one commandment for living things: Survive! And the forms and species and units and groups are armed for survival, fanged for survival, timid for it, fierce for it, clever for it, poisonous for it, intelligent for it. This commandment decrees the death and destruction of myriads of individuals . . . Life has one final end, to be alive; and all the tricks and mechanisms and all the failures, are aimed at that end.

John Steinbeck, *The Log from the Sea of Cortez*

This book is for Martin Birch

Author's Preface

This book is about the behaviour of insects, the most successful animals on Earth. It is based on a very small part of a vast and unique record compiled by photographer-naturalist Ken Preston-Mafham. Many of the subjects have never been photographed before; many of the photographs record new observations. This has made the writing of the book an exciting and challenging experience.

Insects in camera enables us to make much fascinating information available to a wider audience. We hope, therefore, that it will appeal not only to those with a general interest in natural history, but also to undergraduates who may be contemplating entomology or animal behaviour as special topics.

The work of Ken Preston-Mafham explodes many of the myths of natural-history photography. It is simply not necessary for subjects, even as small as insects, to be posed, anaesthetized, in a studio setting. Nor are arc lamps and tripods required. All the photographs in this book were taken with a hand-held camera in the field.

Most picture books in natural history begin with a more or less completed text; only then do the picture researchers comb the agencies for suitable colour transparencies. Inevitably, this results in some variation in photographic style and quality. By contrast, this book began with the pictures. From a collection of 10 500 colour slides, we arrived at a shortlist of about 800. The final selection of 287 slides was made from this. It involved painful decisions at every stage; overall, for every exciting picture we chose, we had to reject 36. And during this period of selection, the shape and scope of this book evolved.

Insects in camera comprises 64 double-page spreads, grouped around seven major themes in the lives of insects. Each theme is the subject of a separate section and has a short introductory essay. I have avoided technical jargon as much as possible. I acknowledge here, the tremendous debt I owe to the hundreds of authors whose published research work cannot be cited directly in the text of a book of this kind. A comprehensive bibliography is given at the end of the book.

Like all branches of science, that of animal behaviour progresses because people ask the right questions. Researchers in the field now pose such questions as 'What are the costs and benefits of behaviour X? How are they juggled to get the best return?' Applied to animal behaviour, the principles of double-entry book-keeping have stimulated many exciting new directions in research. This has coincided with a shake-up in the way behavioural scientists permit themselves to talk and write about animals. Writers such as R. L. Trivers, John Krebs, Nick Davies, and Richard Dawkins talk of insects and other animals taking decisions, making investments, hedging their bets.

This, the 'New Anthropomorphism', has had a liberating effect. It has generated many new insights and the asking of more penetrating questions; it has been a major influence in the writing of this book. If insects do not actually exercise forethought or make investments with the same conscious expectations as their human counterparts, they behave as though they do, driven by the programmed strategies of their genes.

The identification of insects poses many problems and great care has been taken to ensure that the insects illustrated here are correctly named. As far as possible, identifications were based on specimens collected in the field. In some cases I was able to use the resources of the Hope Entomological Collections in the University Museum, Oxford. I also received help from Prof. G. C. Varley, Ivor Lansbury, and Dr Peter Miller. I also thank the following friends and colleagues at the Entomology Department in the British Museum (Natural History) for their help: P. R. Ackery, Barry Bolton, M. J. D. Brendell, Brian Cogan, Mick Day, W. R. Dolling, V. F. Eastop, Mike Fitton, Peter Hammond, W. J. Knight, L. M. Pitkin, John Quinlan, D. R. Ragge, S. Shute, Ken Smith, Colin Vardy, Dick Vane-Wright, and Alan Watson.

I thank Audrey Smith and Ruth Wickett of the Hope Entomological Library, University Museum, Oxford, for much bibliographical help.

I thank my wife, Linda Losito, for her constant encouragement and for her critical reading of the manuscript. I have also benefited from the helpful comments of the following friends and colleagues who have read the text and with whom I have enjoyed stimulating discussions: Drs Cathy Kennedy, Angus McCrae, Peter Miller, Anthony Raw, and Malcolm Scoble. I assume full responsibility for any errors of fact or interpretation. No book, especially this one, can claim to be the last word on its subject. For this reason, I would be pleased to hear from any readers who detect errors or who have alternative interpretations to offer.

We are grateful to our publishers for giving us the freedom to develop the particular format of this book, and we thank the various members of the staff who have worked on the book for showing patience and forbearance above and beyond the call of duty.

I have been sustained throughout by Ken Preston-Mafham's manic enthusiasm for natural history. I thank him for creating the opportunity to write this book and I hope I have done justice to his magnificent photography.

C. O'T.

Oxford, July 1983

Photographer's Preface

The pictures in this book are the culmination of more than ten years of learning how to record the complex behaviour of insects on film. During this time I have been able to make a number of trips to marvel at, and photograph, the rich insect fauna of the tropical world. The success of these overseas ventures would have been impossible without the generous help of a number of people. I particularly thank Dr Gerardo Lamas (Peru), Dr Angus McCrae (Kenya), Dr Jack Price (Trinidad), Ken Heil (USA), Dr John Ismay (New Guinea), Dr Roger Kitching (Australia), Ken Scriven (Malaysia), and Ken Proud (World Wildlife Fund, Java and Borneo).

Not least, I would like to thank Chris O'Toole, whose enthusiasm for my pictures gave me the confidence to meet the challenge of photographing the intricate lives of insects; he has been a constant source of encouragement. I also thank Professor G. C. Varley, who, as Hope Professor of Entomology at Oxford, was the first professional entomologist to give me encouragement; he helped with many of the early identifications.

All the photographs in this book were taken with a hand-held Nikon F2A camera, using a 55 mm Micro-Nikkor lens.

I also owe a debt to my brother Dr Rodney Preston-Mafham and his wife Jean. It is through their help that I have been able to establish the natural history slide library of Premaphotos Wildlife which has enabled me to travel the world over the last few years in search of weird and unusual subjects, not just insects but a whole spectrum of different animals and plants. Without their willingness to stay at home and look after the routine side of running an office while I was away in some exotic tropical location this book would never have been possible.

Finally I should like to thank my parents who have always done their utmost to encourage my interest in photography and whose generosity enabled me to take the pictures in Trinidad which form an integral part of this book.

K. P. M.

Alcester, July 1983

Contents

Classification of Insects

[Orders which are illustrated in this book appear in **bold type.**]

Scientific name	Common name(s)
Class: Insecta	Insects
Sub-class: Apterygota	Wingless insects
Order: Archaeognatha	
Order: Thysanura	Bristletails
Sub-class: Pterygota	**Winged insects**
Order: Odonata	**Dragonflies and damselflies**
Order: Ephemeroptera	**Mayflies**
Order: Blattodea	**Cockroaches**
Order: Isoptera	**Termites**
Order: Mantodea	**Praying mantids**
Order: Zoraptera	
Order: Dermaptera	Earwigs
Order: Grylloblattodea	
Order: Plecoptera	Stoneflies
Order: Orthoptera	**Crickets, grasshoppers, locusts**
Order: Phasmatodea	**Stick insects, walking leaves**
Order: Embioptera	Webspinners
Order: Psocoptera	Booklice
Order: Phthiraptera	Lice
Order: Hemiptera	**True bugs**
Order: Thysanoptera	Thrips, thunderflies
Order: Megaloptera	**Alderflies**
Order: Neuroptera	**Lacewings, ant lions**
Order: Coleoptera	**Beetles**
Order: Strepsiptera	Stylops, twisted-winged insects
Order: Mecoptera	**Scorpionflies**
Order: Siphonaptera	Fleas
Order: Diptera	**True, two-winged flies**
Order: Trichoptera	Caddisflies
Order: Lepidoptera	**Butterflies and moths**
Order: Hymenoptera	**Sawflies, wasps, ants, bees**

I The Life of Insects

1 Success in Diversity

There are more than one million described species of animals; at least 85 per cent of them are insects. One recent estimate puts the total number of insect species at 30 million. This alone testifies to the fact that insects are the world's most successful survivors. It has been calculated that, on average, there are 10 000 million insects per square kilometre of habitable land, or about 200 million individuals for every person living today. Like it or not, we are bathed in a sea of insects.

Insects range in size from species which are smaller than some tiny, single-celled Protozoa, to those which are larger than some of the smallest birds and mammals. This diversity in size is matched by an equal diversity in lifestyles: insects have evolved a bewildering array of responses to the two basic demands of survival, feeding and reproduction.

To say that insects are successful is to say that they have carved out for themselves a tremendous diversity of ecological niches. An animal's niche is the sum total of all the resources it needs—space, time, food, mates — and the particular way that that species acquires and uses them. No two species have identical niches. Some closely-related species may overlap to a greater or lesser extent in their niche requirements, but no two species exploit their niches in exactly the same way; each species is unique. In other words, there are at least 850 thousand to 30 million ways of making a living as an insect.

We can pin-point several major reasons why insects have been able to stake so many of the important ecological claims. First, they are small animals. Creatures the size of cats are excluded from all the specialized livings to be made in crevices under bark or in the soil. Moreover, as a rule, smaller animals have shorter developmental times than larger ones, which means that they have a more rapid turnover in generations. This means, in turn, that they can have a faster rate of adaptation, by natural selection, to new opportunities and challenges.

Secondly, the way insects develop has an impact on diversity. About 88 per cent of known insect species undergo a complete metamorphosis; they pass through larval and pupal stages before becoming adult. The larval stage is devoted entirely to feeding and usually has a diet very different from that of the adult. During the pupal stage, almost all larval tissues are broken down and re-organized to form the new adult, whose major roles will be dispersal and reproduction. Larva and adult are, therefore, separated in space and time by the pupal stage and occupy different niches.

The exploitation of different niches by the same species is not restricted to developmental stages. Adult insects, too, are adept at using what have been called 'fractional niches'. A little solitary bee from the deserts of North America is a good example. *Hoplitis biscutellae* builds brood cells only in the old, persistent nests of mud-dauber wasps. A female separates her cells with partitions made of resin mixed with chewed leaves and flower parts, all collected entirely from the Creosote Bush, *Larrea tridentata*. She forages for pollen and nectar from the same source. Now, there are several other solitary bees which use old mud-dauber nests, but they are not exclusive specialists on *Larrea* for food and building supplies. There are also 22 species which *do* specialize on *Larrea* for food, but only *Hoplitis biscutellae* among them nests in old mud-dauber cells. Insects partition their resources in many similar ways. They have become specialists, so that for any species there is a unique niche or combination of fractional nichès. This, too, contributes greatly to insect diversity.

Another reason for the insect success story is their ability to fly. Many insects, probably far more than have been observed, actively migrate. Dragonflies, hoverflies, moths, and butterflies regularly cross the English Channel and ladybirds and bees have been seen from lightships in the middle of the Baltic. The Monarch Butterfly, *Danaus plexippus*, sometimes crosses the Atlantic between North America and Europe.

Whether as active migrants or as passive, wind-blown members of the aerial plankton, insects have colonized the remotest oceanic islands. Dispersal enables insects to reach new habitats and find new or unoccupied niches. This opens the way to rapid evolution and the appearance of new species. It has been estimated that the 6500 species endemic to Hawaii are descended from just 250 species which managed to cross the Pacific by chance dispersal in the 800 000 years since these volcanic islands emerged from the sea. Insect flight, therefore, makes an important, if indirect, contribution to species diversity.

The success of insects make them important to man. Whether as crop pests or vectors of disease, they cannot be ignored. They are also providers—silk, wax, honey, and an increasing number of pharmacological substances all come from insects. But it is their ecological impact which is of prime importance to man.

Our world would be very different without insects. If they sometimes threaten the very existence of some human communities, we also, paradoxically, depend on them for maintaining ecosystems. Insects are vital in the breakdown and recycling of dead vegetation and animals; they are the sole food of a wide range of vertebrates and are also the world's most important herbivores.

If insects are the major consumers of vegetation, they also maintain much of it. As pollinators of flowering plants, insects not only contribute to the regeneration of Earth's green mantle, but also to much of its visual appeal. And, as predators and parasites of crop pests, they include allies of farmers and growers the world over. Insects, then, occupy key positions; they underpin the complex interrelationships which form the ecological architecture of our planet.

This book is a celebration of the insect success story and of the many beautiful and bizarre ways insects have responded to the twin imperatives of evolutionary success: adapt or perish!

Armour plating: the soft parts of insects are enclosed in a tough, horny exoskeleton of chitin

1. (*Facing*) Larva of a net-veined beetle, *Dulticola* sp. (Fam.: Lycidae), Mt. Kinabulu, Borneo. Resembling a latter-day trilobite, the bark-feeding larva of *Dulticola* shows the armour plating and segmentation typical of insects. The orange spots signal to would-be predators that, like the adult, the larva is highly distasteful and the spines make it difficult to handle. Insect cuticle functions not only as an external skeleton into which muscles are inserted, but also as a waterproof barrier preventing desiccation. The cuticle is made of chitin and protein; it can exist in hard (sclerotized) form or a soft (unsclerotized) state. Hard cuticle has an outer, waterproof layer of wax and is relatively inflexible; soft, pliant cuticle is less waterproof and forms the membranes between body segments, which enable flexibility of movement.

2. (*Below*) A cockchafer beetle, *Melolontha melolontha* (Fam.: Scarabaeidae), England, taking flight. Beetles form the majority of the insect 'armoured divisions'. The forewings are horny and modified as elytra or protective cases for the hind wings, which are folded in a complicated fashion when at rest. Most beetles have a rather lumbering flight while many have lost the ability to fly and have fused elytra. Many desert beetles are flightless and their domed elytra trap a body of humid air over the abdominal respiratory openings, thereby reducing water loss. The cockchafer's lamellate antennae are one of many adaptations by insects to increase the surface area of these olfactory organs, without increasing their length and the risk of damage.

Ever alert, insects are highly sensitive to their surroundings and possess a variety of sense organs

3. (*Facing, top*) A bush cricket or katydid (Fam.: Tettigoniidae) from Trinidad, showing the array of sense organs found on the insect head. The long antennae house organs for the sense of smell; food is tasted and manipulated by two pairs of palps, which may also assess the size of food particles. The short hairs which cover the palps, head, and much of the body are touch receptors — each has a nerve at its base and the slightest pressure on the hair causes it to move and this sends a nerve impulse to the brain. The compound eyes consist of hundreds of facets, each a mini-eye, with its own lens and nerve connection to the brain. Herbivores such as this bush cricket have relatively small eyes.

4. (*Facing, bottom*) Head of the dragonfly, *Aeshna cyanea* (Fam.: Aeshnidae). Dragonflies detect food visually rather than by scent and the eyes are therefore enormous and the antennae relatively small. They are powerful fliers and catch insect prey on the wing. The eyes are very sensitive to movement and abrupt movements can be detected by some species at distances of up to 20 m. Large facets on the upper surface perceive movement and the smaller facets on the lower surface discern the shape of stationary objects such as perches and prey. Although aerial prey accounts for most of a dragonfly's diet, they do feed on resting insects. One species, *Aeshna grandis*, was once seen eating small frogs which it picked up from the ground. Large eyes also enable dragonflies to make the most of low light intensities and many tropical species are crepuscular.

5. A bush cricket or katydid, *Microcentrum* sp. (Fam.: Tettigoniidae), in rain forest, Trinidad, showing the swollen base of the front tibia which contains an 'ear'. The slit opens into a chamber containing a membrane sensitive to airborne vibrations. Male crickets and grasshoppers produce songs characteristic of their species. In crickets, the sound is generated by rubbing together the specially modified, overlapping bases of the forewings. There are different songs, each conveying a particular message — calling, courtship, or territorial aggression. Thus, the communication system comprises transmitters (males) and receivers (males and females).

6. A hoverfly, *Episyrphus balteatus* (Fam.: Syrphidae), lands on burdock. The rapid, darting flight of these aerial acrobats demands a finely tuned integration of sense organs. The large eyes are highly sensitive to movement. Together with the antennae, they help monitor air speed. The hind wings, as in all flies, are reduced to a pair of minute, club-shaped 'halteres', which act as gyroscopic stabilizers, providing the feedback required to maintain precise flight control.

Mouthparts and grooming: supplied with a myriad of sense organs, the mouthparts and body surface must be kept clean

7. A female longicorn or timber beetle, *Prosopocera luteomarmorata* (Fam.: Cerambycidae), Kenya, showing the powerful jaws. Female timber beetles chew holes in wood in which they lay eggs. The larvae, too, have strong jaws and burrow through the wood on which they feed. Wood is relatively indigestible and is low in nutrients, so larvae may take anything from 3 to 32 years to complete development. The adult beetles of both sexes use their jaws to make an emergence hole in the timber in which they have developed. There are 20 000 species in the Cerambycidae, a diverse and colourful family, some of which are the largest of all insects.

8. A female weevil, *Rhinastus latesternus* (Fam.: Curculionidae) from Peru. Weevils, like all beetles, have biting mouthparts. Their jaws, however, are at the end of a snout-like prolongation of the head called the rostrum. There is much variation in rostrum length throughout the family. It is very long in species where the females use it as a boring device to make holes in seeds or nuts in which to lay eggs. The rostrum is therefore proportionately longer in the female than in the male. With more than 60 000 species, the Curculionidae is larger than any other family of animals.

9. Bugs have piercing and sucking mouthparts. Here, a squashbug (Fam.: Coreidae), Trinidad, grooms its tubular mouthparts with its front legs. Grooming activity is universal throughout the insects and keeps mouthparts and sense organs such as antennae and eyes, free from debris. The movements involved are stereotyped and characteristic of the group to which the insect belongs. Much time is spent in grooming — this bug spent 15 minutes just cleaning its mouthparts.

10. (*Above*) A Red-headed Cardinal Beetle, *Pyrochroa serraticornis* (Fam.: Pyrochroidae), England, grooming its serrated antennae with its front legs and maxillary palps after feeding on the nectar of Charlock. The Pyrochroidae is a small family, with little more than 100 species, mainly in the Northern Hemisphere. The larvae live under bark, where the adults are also often found.

11. A praying mantis, *Acontista* sp. (Fam.: Hymenopodidae), Trinidad, using her mouthparts to clean the tip of her abdomen. This contortion is possible because the overlapping plates of the abdomen are joined by flexible, intersegmental membranes. She groomed non-stop for 15 minutes. Perhaps she had just laid a batch of eggs; these are protected by a frothy secretion which hardens to form a spongy mass.

Ever sensitive, insects respond to a changing environment, be it with an appropriate posture, or opportunistic feeding

12. A male dragonfly, *Trithemis ellenbecki* (Fam.: Libellulidae), Masai-Mara Game Reserve, Kenya, responds to the heat of the midday sun by adopting the so-called 'obelisk' posture: he tilts his abdomen upwards so that it points straight at the sun, presenting the minimum amount of body surface to the sun's rays. The wings shade the thorax and the risk of overheating is minimized. In the early morning, when it is cooler, the dragonfly adopts a different posture, with the wings spread wide and abdomen flat against a stone or log, maximizing the amount of body surface exposed to the direct rays of the sun.

13. The males of many solitary bees and wasps sleep communally on twigs and branches. Here, a trio of male bees, *Amegilla* sp. (Fam.: Anthophoridae) roosts on dead stems in open *Acacia* woodland, Kenya. Both sexes foraged in large numbers at labiate flowers during the day. After a heavy and sudden evening thunderstorm, the males aggregated for sleep, while the females sheltered in their nests in the ground. By spacing themselves out on the ends of dead stems, the males ensure they will rapidly warm up with the next morning's sun and be able to resume their relentless search for nubile females.

14. Extreme conditions test the opportunism of insects. In the severe drought which afflicted Europe in the summer of 1976, many insects quenched their thirst in unusual ways. Here, a worker of a social wasp (yellowjacket), *Vespula germanica* (Fam.: Vespidae), England, used its powerful jaws to bite off the flower head of a thistle; it drank the sap oozing from the cut stem.

15. After the wasp had drunk her fill, she was followed by a succession of thirsty insects, including soldier- and leafbeetles. Here, a quartet of ladybirds, *Coccinella septempunctata* (Fam.: Coccinellidae) takes advantage of the wasp's action. They, like the other drinkers, have weaker jaws than the wasp and could not have tapped such a rich source of moisture by their own efforts.

16. Grasshoppers from two families, Eumastacidae (left) and Acrididae (right), feeding on human excrement, Peru. Leaves are the normal food of grasshoppers. However, in the vicinity of paths through the rain forest near Tingo Maria, human excrement is widely available for any insect opportunists which can use it. The two grasshopper species shown here were common throughout the area and were found feeding only on human excreta, suggesting that it is a locally important food.

17. Short-tongued bees often cheat on the normal bee–flower relationship. They bite their way through the rear of long-tubed flowers to reach the nectaries, thereby robbing the flower, without contacting the pollen-bearing male parts (anthers) or the pollen-receptive female part (stigma). Here, a honeybee worker, *Apis mellifera*, robs a bluebell. Instead of biting her way in, though, she prizes the petal bases apart with her jaws and tongue. Her tongue is, however, long enough to reach the nectaries via the legitimate route. Perhaps, in this case, robbing gives higher nectar rewards.

Growth and development: incomplete and complete metamorphosis, two ways of growing up

18. A newly adult grasshopper, *Taeniopoda auricornis* (Fam.: Acrididae), Mexico, hangs from its nymphal skin. Grasshoppers have an incomplete metamorphosis: the egg produces a small nymph which lacks wings and functional gonads, but nevertheless resembles a miniature adult. Insects with this sort of development are called 'exopterygotes' because their wings develop externally. Nymphs usually have the same diet and habitat as the adults. Encased in an exoskeleton, they cannot grow at a steady rate. Instead, the nymph moults several times and growth occurs while the new exoskeleton is still soft. With each successive moult, more adult characteristics appear.

19. A nymph of the dragonfly, *Libellula quadrimaculata* (Fam.: Libellulidae) has climbed on to a rush prior to emergence as an adult. The nymphal skin splits first in the head region and the young dragonfly eases its way out. The nymph spends 2 to 3 years as a voracious aquatic predator, eating other insects, tadpoles, and small fish. It seizes prey with a remarkable structure called the 'mask', a hinged modification of the mouthparts, which rapidly extends and impales prey on the hook-like palps at its tip. The mask then retracts, bringing the prey within reach of the jaws.

20. The newly-emerged dragonfly rests on its nymphal skin, and expands its wings by pumping blood into the complex network of veins. For a day or so, the wings retain their milky cloudiness, the body pigments remain pale, and the insect flies little. There is often a mass exit of nymphs from the water and they and the newly-emerged, teneral adults, are vulnerable to predation by birds and take up sheltered positions. Those that failed to do so around this Gloucestershire pond fell prey to a pair of Spotted Flycatchers which nested nearby.

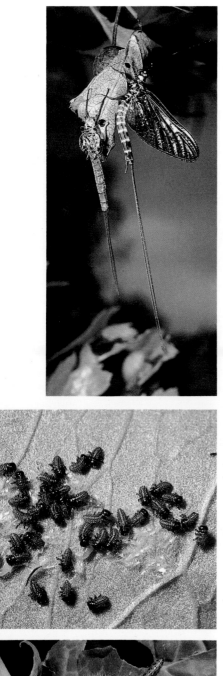

21. A newly-hatched adult mayfly, *Emphemera danica* (Fam.: Ephemeridae) rests next to the cast skin of the last nymphal stage. The latter is called a 'subimago' because, unlike the aquatic nymphs of other exopterygotes, those of mayflies resemble the adults (imagos) in their ability to fly. The nymph moults to produce the subimago under water, or at the surface, or on a stone or plant at the water's edge. The wings of the subimago are opaque, fringed with hairs, and the legs and other appendages are not of the full length. It flies to some vantage point, where the final moult takes place and the fully-formed adult emerges. The adults live for a few hours only, have no mouthparts and do not feed. The loss of mouthparts begins in late nymphal life and is complete at the subimago stage. Most of the 'flies' of fly-fisherman are modelled on mayflies.

22. Newly-hatched larvae of a leaf-beetle, *Gastrophysa viridula* (Fam.: Chrysomelidae) eat their eggshells on the foodplant, Wood Dock. Beetles, like all higher insects, have a complete metamorphosis: the larvae bear no resemblance to the adults and often differ from them in diet and habitat. There are several larval moults and larva and adult are separated by a pupal stage, during which larval tissues are broken down and reorganized into adult form. Insects with complete metamorphosis are called 'endopterygotes' because their wings develop internally.

23. Caterpillars of the Cabbage White Butterfly, *Pieris brassicae* (Fam.: Pieridae), England, feeding on cabbage. The bright colours signal their distasteful nature to birds (see 218). Warningly-coloured larvae are often communal, thereby increasing the effectiveness of their coloration. The larval stage is a time of voracious feeding, and communal feeders are often major pests of crops; they can cause much damage in a short time.

2 Food and Feeding

A swarm of the African Desert Locust, *Schistocerca gregaria*, may extend over hundreds of square kilometres and contain, perhaps, 40 000 million individuals and weigh 70 000 tons. Such a swarm will eat nearly four times as much food per day as the populations of Greater London or New York City.

The locust swarm symbolizes the prodigious consumption of insect herbivores and their ability to process rapidly large amounts of green leaves. And it exemplifies one of the many remarkable adaptations of insects for finding their way to food.

A migratory swarm is a response to diminishing food sources and is usually stimulated by drought. The population density of locusts increases as their food plants die back; they find themselves stranded on ever-dwindling islands of vegetation.

The locusts take to the air *en masse* when a cold front approaches and a swarm is born. It flies with the front on its broad, counter-clockwise sweep over the landscape. A cold front eventually brings rain and the locusts are on hand to eat the new flush of green vegetation and resume breeding. They are therefore adapted to exploit regional weather systems, in their quest for food.

Very few insects, however, are as dramatic and conspicuous as the Desert Locust. But many thousands of unobtrusive species are just as finely tuned to the challenge of finding food. Some, indeed, are bizarre.

Sometimes, the absence of a particular specialist enables other insects to take on new and unexpected diets. In Hawaii, there are no native praying mantids, so the niche for surprise-attack predatory insects was vacant. It has been filled by caterpillars of several species of *Eupithecia*, a genus of geometrid moths. Like many geometrid caterpillars, each resembles a short twig and remains motionless during the day, attached to a branch or leaf by its rear end. But there the similarity ends. When a small insect inadvertently contacts specialized sensory hairs, it triggers off a lightning response: the caterpillar seizes the insect with its well-developed legs, resumes its normal posture, and eats the prey.

Several species of moths have abandoned their normal diet of nectar. Instead, they use their sucking mouthparts to feed on the body fluids of animals. In the Far East, *Lobocraspis griseifusa* drinks the protein-rich tears of buffalo and cattle. Another oriental moth, *Calpe eustrigata*, has gone one stage further and pierces the skin of mammals and sucks their blood; in an ecological sense, it has become a mosquito.

More than half of all insect species, however, are plant feeders. And the patterns of insect herbivory are largely determined by the chemical warfare waged by plants. Plants produce a variety of weapons, the so-called secondary compounds. These include tannins, which bind with proteins and make them indigestible, and a variety of highly poisonous substances such as cyanide and alkaloids. These deter most grazing animals and insects. But some insects can not only detoxify the poisons, but also store them in their tissues and use them to their own advantage for defence or sexual signalling.

The capacity of insects to cope with chemical warfare is demonstrated by the many species which have evolved resistance to DDT within 40 years of the widespread introduction of this insecticide.

Insect specialists are everywhere. Beetles and flies are vital in pastureland for disposing of cattle dung. Without them, the areas of available grass would soon be reduced. For this reason, African species of dungbeetles were introduced into Australia, where the native beetle fauna was unable to cope with the amount of dung produced by the vast herds of introduced cattle. The African beetles evolved with a diverse fauna of grazing animals. Transplanted to Australia, they now also control pestiferous bushflies which breed in the dung.

The specialized feeding behaviour of insects is important to man in other ways. On a world basis, nectar-seeking bees pollinate crops to the value of 1770 million US dollars per annum. It was possible to enjoy cheap lamb and dairy produce from New Zealand because of the pollination services of a few species of European bumblebees, introduced there in the late nineteenth century. The native bees are poorly adapted to clover flowers, imported as a forage crop for livestock. Within 5 years of the introduction of bumblebees, New Zealand changed from being an importer of clover seeds to being an exporter.

Although diversity through specialization is a recurring theme of this book, some insects are remarkable for the wide range of apparently inedible foods they can exploit. We can only marvel at the larvae of the little scuttle fly, *Megaselia scalaris*. They feed on almost any decaying matter, but have also been reared from emulsion paint, shoe polish, human cadavers pickled in formalin, and the lungs of living Japanese students. If there is a living to be made, somehow, somewhere, there is an insect which can do it.

Biting and chewing mouthparts: herbivory in the Orthoptera, grasshoppers and crickets

24. A grasshopper, *Chromacris colorata* (Fam.: Acrididae), Mexico, feeds on a poisonous plant of the potato family (Solanaceae). Grasshoppers straddle both sides of a leaf while eating, so that their jaws gain direct purchase on the leaf margin, like scissors on paper. This species sequesters poisonous alkaloids from the plant, a fact which is advertised by the warning coloration sported by both nymphs and adults, which feed gregariously.

25. (*Above*) A female grasshopper, *Taeniopoda auricornis* (Fam.: Acrididae), Mexico, eats a *Mimosa* flower. Grasshoppers often eat flower parts. The petals are usually softer than leaves and the anthers yield protein-rich pollen, while the bases of petals often bear nectaries, which provide an energy-packed food. Flower-feeding Orthoptera are often pests where flowers are grown commercially in the tropics.

26. (*Facing, top*) A final-instar bush cricket or katydid nymph, *Microcentrum* sp. (Fam.: Tettigoniidae), eats the petals of an *Hibiscus* flower in Trinidad. Note that one of the hind legs is braced against an adjacent leaf, freeing the front legs to handle a large piece of petal while smaller pieces are manipulated by the palps and passed to the jaws.

27. (*Facing, bottom*) A cricket nymph, *Nisitrus* sp. (Fam.: Gryllidae), Mt. Kinabulu, Borneo, eating the anthers of a cultivated flower. Gryllids are omnivorous and will eat a wide range of animal and plant foods. In some parts of the world, they are used as living corn removers because their powerful jaws are adept at removing horny skin. Some peoples keep the singing adults as pets.

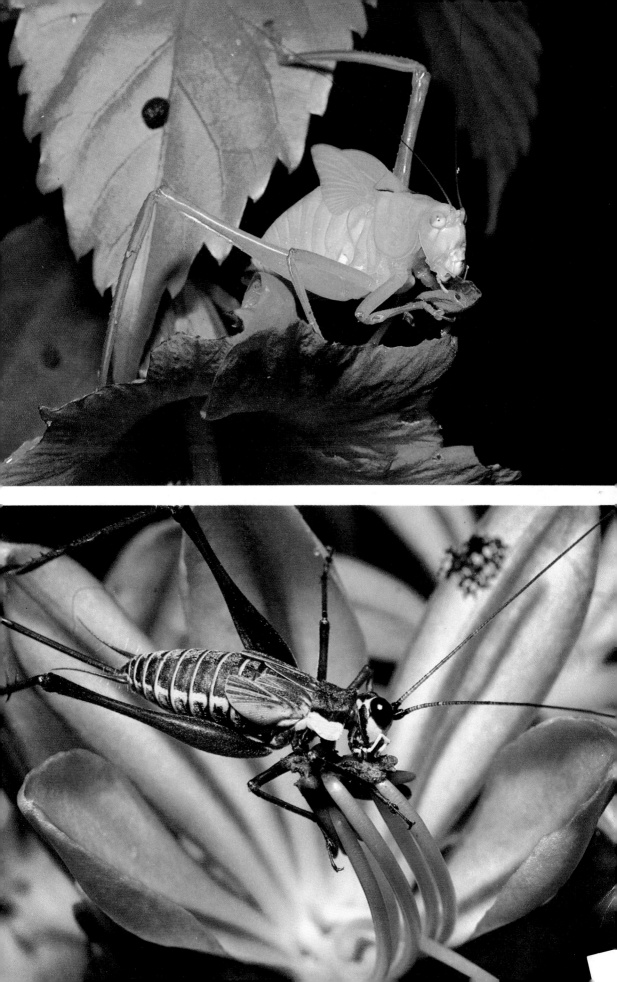

Biting and chewing mouthparts: herbivory in beetles and primitive moths

28. Leaf beetles, *Chrysomela populi* (Fam.: Chrysomelidae), eating the leaves of Creeping Willow, *Salix repens*, growing on dunes. Adult leaf beetles are often brightly coloured. They bite leaf margins, producing a scalloped effect. Many are highly specific in their choice of food plants and are pests of crops. Some transmit bacterial and viral diseases of plants. The gregarious larvae skeletonize leaves by eating the outer layer of cells.

29. A weevil, *Brachyomus octotuberculatus* (Fam.: Curculionidae), Trinidad, straddles a leaf while eating the margin. This posture is typical of weevils with a short rostrum. Those with a longer rostrum drill circular holes in the leaf blade. Many weevils are specialist pests of crops, including cotton, bananas, and broad beans. *Sitophilus granarius* is a cosmopolitan pest of stored grains.

30. A quintet of small moths, *Micropteryx calthella* (Fam.: Micropterigidae) feeding on buttercup pollen. Moths of this primitive family have functional biting mouthparts rather than the sucking proboscis of the higher Lepidoptera. They grind pollen grains with their jaws. The larvae feed on mosses, a fact which emphasizes the antiquity of the family — as a general rule, primitive herbivorous insects have primitive foodplants.

31. (*Facing, top*) A caterpillar of the Yellow Tail Moth, *Euproctis similis* (Fam.: Lymantriidae), eats the petals of Dog Rose, *Rosa canina*. Other foodplants of this species include sallow, hawthorn, birch, beech, and oak, as well as fruit trees such as apple and pear. The larvae of all Lepidoptera have tough jaws and many gregarious feeders can process large amounts of leaf material with remarkable rapidity, sometimes causing serious defoliation.

32. (*Facing, bottom*) A timber or longhorn beetle, *Rhagium bifasciatum* (Fam.: Cerambycidae), eating the pollen of Dog Rose, England. Timber beetles are regular visitors to flowers in high summer, where they chew the anthers to liberate pollen. Flowers are also meeting places for the sexes and many species can be seen mating at blossoms.

Biting and chewing mouthparts: active predators with acute vision and rapid reactions

33. (*Facing, top*) The nymph of a flower-mimicking praying mantis, *Pseudocreobotra ocellata* (Fam.: Hymenopodidae), eating a bee-fly, Kenya. Praying mantids are so called because they sit with their front legs held out in an attitude of prayer. They react with extreme rapidity when an insect comes within reach of the heavily-spined, raptorial front legs. In this case, the prey is one of the fastest insect fliers, which is a tribute to the mantid's lightning response.

34. (*Facing, bottom*) A male tiger beetle, *Cicindela aurulenta* (Fam.: Carabidae), Borneo, mounts a female. His powerful, multi-toothed jaws are used here to grip the female's body. Tiger beetles are fierce predators and will attack any suitably-sized insect. They fly readily and are the fastest insect runners. The larvae are carnivores, too, and live in vertical burrows in the ground; they dart out and seize any unsuspecting insects which wander by.

35. Dragonflies are superbly adapted for catching insect prey on the wing. Here, a male *Orthetrum julia* (Fam.: Libellulidae), Kenya, eats an insect. The geometry of the thorax is modified so that the legs tilt forward in a grasping position. This is enhanced by the spines on the legs, which act as an aerial grab as the dragonfly hawks along a regular beat in search of prey.

36. Damselflies have the same adaptations for aerial predation. Here, a male *Enallagma cyathigerum* (Fam.: Coenagriidae), a damselfly common in Britain and Europe, eats a caddisfly. Together with the dragonflies, the damselflies comprise the Order Odonata. They are less powerful fliers than dragonflies, with fluttering rather than hawking flight and do not wander so far away from water; they are, nevertheless, just as efficient as predators.

37. An assassin bug (Fam.: Reduviidae) falls prey to a female praying mantis, *Parasphendale agrionina* (Fam.: Mantidae), in the Shimba Hills, Kenya. Assassin bugs are themselves hunters and killers. With birds and mammals it is very unusual for predator to eat predator, but in the opportunistic world of mantids anything goes. Mantids sometimes eat small mammals and in Queensland, Australia, tree frogs are regularly eaten.

Sucking and piercing mouthparts: Herbivores. Some groups have independently evolved tubular mouthparts for imbibing liquid food

38. Butterflies and moths have a very long tubular proboscis and feed mainly on nectar. The proboscis is tightly coiled under the head when not in use. Extension is brought about not by direct musculature but by the pumping of blood (haemolymph) down channels in the proboscis. Here, a Small Tortoiseshell Butterfly, *Aglais urticae* (Fam.: Nymphalidae) probes the flowers of a garden stonecrop (*Sedum* sp.).

39. Flies, too, have sucking mouthparts, though they are controlled by direct musculature. Here, a hoverfly, *Rhingia campestris* (Fam.: Syrphidae) feeds on the nectar of *Geranium pratense*. The fly is in effect robbing the flower because it is too small to brush against the pollen-bearing anthers or the female stigma and bring about pollination. *Geranium* spp. are normally pollinated by bees, especially bumblebees.

40. Sucking mouthparts are central to the bee–flower relationship. Flowers provide nectar and pollen as a reward for pollinating insects. Bees are the most important pollinators and nectar is an energy-rich food which fuels their foraging flights. They convert it to honey for storage and for larval food. There are many highly specific relationships between flowers and bees. Many groups of bees have evolved long tongues, which enable them to exploit the nectar of long-tubed flowers. The deep-tubed flowers and long-tongued bees evolved together, each becoming better adapted to the anatomy of the other, so that the plants attract a guild of specialist pollinators and ensure reproductive success, while the bees enjoy a guaranteed and exclusive food source. Here, a bumblebee, *Bombus pascuorum* (Fam.: Apidae) probes a scabious flower with its tongue.

41. Aphids or greenfly tap the sap-conducting vessels (phloem) of plants. They do not actively suck the sap but rely passively on the fluid pressure exerted by the plant. This is so great that the sugary sap passes through the gut and appears as droplets of honeydew at the anus. Here, a group of *Macrosiphum cholodkovskyi* feeds on Meadowsweet. One of them gives birth to live young.

42. A shield-backed bug, *Tectocoris diophthalmus* (Fam.: Scutelleridae), sucking juices from an *Hibiscus* fruit, Australia. The mouthparts of these bugs, and their relatives, the shieldbugs (Pentatomidae) are powerful and can penetrate tough fruits. Saliva, containing enzymes, is pumped down the rostrum into the plant tissues and some pre-digestion takes place before the muscular suction pump in the head draws up the juices.

43. Cicadas are among the largest of plant-sucking insects. Best known for the sound-producing abilities of the males, they are mainly tropical, with more than 1600 described species. The adults are usually associated with trees and their mouthparts can penetrate bark. Here, a tree-feeding *Brevisiana* sp. sucks sap from a tree in Kenya. The nymphs are subterranean and suck sap from tree roots.

Sucking mouthparts: butterflies; they cope with a wide variety of liquid and semi-liquid foods

44. The males of many tropical butterflies congregate and feed at urine-soaked ground. Here, in Malaysia, a cloud of 'whites' (Fam.: Pieridae) and a lone *Graphium* sp. (Fam.: Papilionidae), feed at urine provided by the photographer. The butterflies use the urine to maintain the correct balance of salts in their body fluids, always a problem in the tropics, and also to replace those lost in sperm packages (spermatophores) during mating (see also 46).

45. A Small Postman butterfly, *Heliconius erato* (Nymphalidae: Heliconiinae), Trinidad, feeds at a flower. *Heliconius* collect pollen and compact it into a ball on the proboscis. Regurgitated nectar is mixed with this and the pollen is agitated for several hours, then discarded. The nectar is re-imbibed, together with proteins released by the crushed pollen grains. This unusual diet enables these butterflies to live in excess of six months.

46. A male of an African butterfly, *Charaxes pollux* (Fam.: Nymphalidae) feeds on mongoose dung in the Kakamega Forest, Kenya. This and other nymphalids often feed on carnivore dung — leopard droppings are said to be most attractive — and are believed to exploit this rich source of amino acids and salts (see 44). Neither sex of *Charaxes* spp. feeds at flowers; males and females feed at fermenting fruit and sap. Only the males visit dung, and this may indicate that the dung provides some chemicals necessary for sexual signalling. Note that the butterfly is joined by a fly and several stingless-bee workers (*Trigona* spp.), which also feed on the dung.

47. (*Facing*) Damp ground at the margins of puddles and streams also attracts butterflies. Here, *Philaethria dido dido* (Nymphalidae: Heliconiinae) drinks moisture from drying mud in Peru. Presumably the water contains dissolved salts which the butterfly needs. The unusual longevity of the exclusively neotropical Heliconiinae is made possible not only by proteins derived from pollen, as in *Heliconius* (see 45), but also from nectars rich in free amino acids. These are secreted by plants adapted for pollination by heliconiines.

Sucking and biting: a miscellany of sometimes unusual foods

48. A hoverfly, *Episyrphus balteatus* (Fam.: Syrphidae) eats aphid honeydew on a leaf. The apex of the tongue comprises a broad pad called the labellum. The underside of this has a system of food channels called pseudotracheae, through which the fly sucks liquid food. The labellum can be reflexed to expose the mouth and allow solid food particles to be ingested. Hoverflies eat pollen grains in this way.

49. A large mosquito, *Toxorynchites* sp. (Fam.: Culicidae), Trinidad, sucks nectar at a flower. The large, aquatic larva of *Toxorynchites* is carnivorous and accumulates enough protein to satisfy the needs of the adult, both sexes of which feed only on nectar. This is unusual, because the females of most mosquitoes require the protein in blood meals from bird or mammal hosts in order to produce eggs.

50. A mosquito, *Malaya* sp. (Fam.: Culicidae), solicits and 'robs' honeydew from a predatory ant, *Ectatomma* sp. (Fam.: Formicidae), Trinidad. The mosquito apparently taps the ant's antennae with its long front legs and the ant responds by regurgitating a drop of food. In the nest, worker ants solicit food from returning foragers by tapping their antennae with their own. It seems, therefore, that species of *Malaya* have 'broken' the ants' food-sharing 'code'.

51. (*Facing, top*) A mêlée of seven scarabaeid beetles of three species jostles for position at a fermenting sap flow on a tree in Kenya. The two largest beetles, a female (*left*) and male (*right*) are a pair of *Chelorrhina polyphemus*. The male may have started the sap flow by wounding the tree with the horns on his head, though their main function is in combat with rival males (see 68). The sap flow attracts females and a succession of other insects.

52. (*Facing, bottom*) Females of a large ground-nesting solitary bee, *Epicharis* sp. (Fam.: Anthophoridae), drinking water on a mud-flat in Peru. *Epicharis* is a neotropical genus of rapid-flying bees. The females have modified front feet which form a spoon-like device for scooping up oils secreted by certain flowers. The bees mix the oil with pollen for larval food. They forage over distances of 25 km or more and must replenish their body fluids regularly.

Piercing and sucking mouthparts: predators. Many insects have made the transition from sap to animal body fluids

53. An assassin bug, *Apiomerus flaviventris* (Fam.: Reduviidae) sucks the blood of a solitary bee (Fam.: Anthophoridae) in Mexico. Reduviids often lie in wait at flowers for unsuspecting prey. The bug injects saliva which paralyses and kills the victim and breaks down its tissues. It then sucks up the resultant protein-rich liquid. Assassin bugs are often seen carrying their prey around, impaled on the rostrum.

54. A female damselfly, *Pyrrhosoma nymphula*, being eaten by a trio of pondskaters, *Gerris lacustris* (Fam.: Gerridae) in England. The damselfly came to lay eggs in the pond and was attacked and overcome, although defended by her mate (see 125). Pondskaters are extremely sensitive to vibrations on the surface film of water and usually feed on struggling, drowning insects which have fallen into the pond.

55. A predatory shieldbug, *Euthyrynchus floridanus* (Fam.: Pentatomidae) feeds on a leaf-beetle, *Leptinotarsa undecemlineata* (Fam.: Chrysomelidae), Mexico. The prey's warning coloration signals its poisonous nature, but the poison and striped livery are only effective with vertebrate predators such as lizards and birds. The predator is itself warningly coloured (aposematic), which advertises the repugnant smell produced in the thorax; pentatomids are often called 'stink bugs'.

56. (*Facing, top*) The Common Yellow Dungfly of Britain and Europe, *Scathophaga stercoraria* (Fam.: Scathophagidae), sucks the blood of a hoverfly. Dungflies hunt and mate at cow dung, where the females also lay eggs (see 124); the larvae eat dung. Adults also hunt at flowers, where flies and small bees are their victims. The prey is pierced in the neck and ceases to struggle immediately, although there is no evidence that any poison is injected.

57. (*Facing, bottom*) A robberfly, *Microstylum* sp. (Fam.: Asilidae), feeds on a beetle, Kenya. Asilids are rapid fliers and catch insect prey on the wing. They have binocular vision and the head is very mobile: they turn their heads continuously, watching for prey while resting on some vantage point. An individual often patrols a regular area. The powerful, spiny legs are adapted for grasping; the rigid proboscis is very sharp and the prey is reduced to a dry husk.

Scavenging: dead and dying insects are a valuable food resource for a wide range of insects

58. A scavenger fly, *Tetanocera hyalipennis* (Fam.: Sciomyzidae), England, feeding on the remains of a dead fly discarded by a spider from its web – note the silk threads binding the corpse. Sciomyzids are general scavengers in damp, marshy places.

59. Scorpionflies (Mecoptera) eat dead and dying insects. Here, the common European species, *Panorpa communis* (Fam.: Panorpidae), eats the remains of a mayfly in a spider's web. This species often scavenges on insects trapped in this way and seems able to do so without attracting the attention of the spider. However, if one accidentally blunders into a web, it is usually eaten by the alerted spider.

60. Opportunistic feeding by a scorpionfly: a male *Panorpa communis* eats an injured damselfly, *Enallagma cyathigerum*. Damselflies are predators (see 36) and normally would be too large for a scorpionfly to tackle, but injured individuals are fair game. Note the beak-like front of the scorpionfly's head. The name refers to the scorpion-like way in which the male bears the tip of the abdomen.

61. (*Facing, top*) Here a female *Panorpa communis* feeds on the semi-liquid remains of a dungfly, *Scathophaga stercoraria*, which has been killed by an entomophagous (= insect-eating) fungus. This is the commonest form of scavenging by scorpionflies. They are immune to the infection because the fungal spores germinate only in the specific fly host (see 64).

62. (*Facing, bottom*) Too small to catch their own prey, flies of the family Milichiidae are the ultimate scavengers. They associate with spiders, assassin bugs (Reduviidae), and robberflies (Asilidae), feeding on body fluids exuding from the wounds in their prey. The flies are too small to be threatened by the larger predators. Here, three milichiids, *Desmometopa* sp., share the blood of a honeybee caught by a robberfly, *Promachus* sp. This kind of relationship is called commensalism.

The tables turned: insects exploited by mites, fungi, and spiders

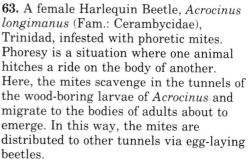

63. A female Harlequin Beetle, *Acrocinus longimanus* (Fam.: Cerambycidae), Trinidad, infested with phoretic mites. Phoresy is a situation where one animal hitches a ride on the body of another. Here, the mites scavenge in the tunnels of the wood-boring larvae of *Acrocinus* and migrate to the bodies of adults about to emerge. In this way, the mites are distributed to other tunnels via egg-laying beetles.

64. Hoverflies, *Platycheirus peltatus* (Fam.: Syrphidae), killed by an entomophagous (= insect-eating) fungus, *Entomophthora musci*. The spore-bearing bodies of the fungus are clearly visible as a white fur bursting through the intersegmental membranes of the abdomen. It is not clear why the infected flies should die in congregations. One suggestion, based on observations of fungi which attack other types of insect, is that a weakened, infected fly settles on a grass flower and dies. The body is attractive to other flies of the same species, which feed on the partially-digested corpse and ingest infective spores. They fly off and deposit spores on other flowers before they themselves die. Other flies visit the same flowers or the corpse, pick up the spores and the disease spreads rapidly through a local population. The important point is that the fungus manipulates the feeding behaviour of healthy flies to ensure its propagation.

65. (*Facing*) It is not surprising that many predators of insects have learned that flowers attract a ready source of food. Many of them, therefore, spend their lives lurking on, or under, or in flowers. Here, a crab spider, *Misumena vatia* (Fam.: Thomisidae) eats a worker honeybee which visited the white flowers of Ox-eye Daisy. The spider waits motionless, with its two front pairs of legs raised. Its pure white camouflage renders it virtually invisible to potential prey. Some individuals are sulphur yellow and live on flowers of the same colour. *Misumena* can change its colour from white to yellow and vice versa, over 2 or 3 days. The yellow colour is caused by fluid in the superficial cells of the cuticle. This can be removed, so that the transparent layer of cuticle appears white because of an inner layer of guanin crystals. The camouflage is so effective that the easiest way to find the spider is to look for the accumulated remains of dead prey hanging down under the flower.

3 Finding a Mate

Insects are in business to make more insects. They have evolved an astonishing variety of means to this end, including behaviour we normally associate with birds and mammals — territoriality, lek displays, courtship gifts, and trial-of-strength contests; all of these are described in the next few pages.

With many hundreds, or even thousands of species seeking mates in a given locality, it is vital that the sexes attract partners of the right species. They must announce their presence in clear and specific terms. Insects use a variety of calling cards, including scents, song, visual signals, and gestures. Usually, one sex is the signal transmitter, and the other the signal receiver.

Mate attraction by scents (sex pheromones) is widespread in insects and is known in groups as diverse as flies, beetles, wasps, and bees, but nowhere is it more refined than in the moths and butterflies. A male silk moth, *Bombyx mori*, can detect the scent of a calling, nubile female at a distance of 4.4 km (2.7 miles) and orientate to it. This is a fantastic achievement when one considers that the specific silkmoth scent, bombykol, is present in only minute quantities; it forms less than one millionth of a female's body weight. The dilution over several kilometres is immense, yet this system works because it requires only a few hundred scent molecules to reach a male's feathery antenna for it to trigger a behavioural response. Our own sensory world is biased in favour of sight and sound and it is impossible for us to comprehend the sensitivities involved in insect communication by scent.

In many insects, mate recognition is by visual cues, as in butterflies, where males recognize females by their conspicuous wing patterns. Only then does scent play a part. The male butterfly may court his mate with a display dance, at the same time drenching her with an aphrodisiac scent, the components of which he may have obtained from withered vegetation.

We are used to the idea of courtship being a matter of boy meets girl, but with some insects, it also involves boy meets boy. The brilliantly metallic males of the South American euglossine bees collect scents from the orchids they pollinate and store them in their expanded hind legs. These fast-flying bees use the scents not to attract females, but to attract other males. They then set up a display aggregation and it is this conspicuous 'lek' which attracts females.

Courtship by song is well known and is also universal in cicadas, crickets, and grasshoppers. In these insects, the male is usually the calling sex, a fact noted by the Greek satirist, who wrote "Happy the cicadas' lives, for they all have silent wives". Sometimes a cricket species has more than one song. Females are attracted by a 'calling' song and then courted by a 'love' song, which may or may not have the desired effect. Here, the females have a choice.

Female choice is found, too, in insects where males stage visual displays or leks, or offer courtship gifts in the form of pre-packaged insect prey. The male with the most vehement song, best display or gift, is likely to have the best genes and is therefore the most suitable individual to father a female's offspring.

Courtship, then, may involve scent, sound, visual cues, or a combination of these. But this advertising behaviour is sometimes not without additional cost. The enemies of insects may latch on to the specific signals transmitted by an insect and act accordingly. A parasitic fly in North America has learned to recognize the courtship song of the cricket, *Gryllus integer* and uses it to locate hosts on which it lays its eggs; the specific egg parasites of moths may also home in on the sex pheromones released by host females.

The genitalia of insects, especially of males, are usually highly characteristic of the species, so much so that entomologists use them as a kind of fingerprint for distinguishing similar species. It used to be thought that the high specificity of genitalia, with their complex systems of hinged levers, hooks, and eversible sacs, was a kind of 'lock-and-key' mechanism to prevent mating between different species. Now it seems likely that the differences between the genitalia of different species are not sufficient to be a barrier to sperm transfer. It is more realistic to regard the highly characteristic genitalia of an insect as the last in a sequence of specific signals for mate recognition.

Each stage in the courtship sequence is a check, ensuring the identity of the pair, so that time and energy are not wasted in fruitless pairings. The genetic endowment of each individual is worth protecting in this way: hybridization, if it occurs, almost always results in sterile offspring.

The particular ways in which a species makes a living, seeks and attracts mates, are all part of a survival kit, packaged and orchestrated by the genes. And the genes are geared to immortality by making copies of themselves via sexual reproduction.

Insects with no apparent courtship: machismo and rape strategies in males

66. Mass matings of Soldier Beetles, *Rhagonycha fulva* (Fam.: Cantharidae), are a familiar sight on flowers in Europe. They are predatory, though the females graze the anthers of flowers during mating, a process which lasts a long time and which is apparently initiated by males mounting any available females. Their name harks back to the days when British Army uniforms were red.

67. This mating pair of weevils, *Otiorhynchus singularis* (Fam.: Curculionidae), England, began copulation after the male made persistent attempts to overcome the resistance of the female, which she expressed by trying to dislodge the male with her hind legs. 'Playing hard to get' is a strategy adopted by many female animals. It ensures that only the strongest and most persistent suitors succeed in mating with them.

68. This female chafer beetle, *Chelorrhina polyphemus* (Fam.: Scarabaeidae; subfam.: Cetoniinae), Kenya, also played hard to get, but sheer brute force by the male prevailed. Both sexes are attracted to fermenting tree sap oozing from wounds (see 51). Sometimes the males use their horns to make wounds in order to attract females. They also use their horns in trial-of-strength contests with competing males.

69. (*Facing*) Courtship and finesse are absent in the dragonfly *Sympetrum striolatum* (Fam.: Libellulidae), shown here mating. A male will quieten a struggling female by bending his wings back through 90° to slap her repeatedly on either side of her head. This pair are in the so-called 'wheel' position, a contortion dictated by the male's peculiar sexual anatomy. His genital opening is in the normal position at the end of the abdomen, between a pair of claspers. But, before finding a mate, he transfers sperm to the front of his abdomen, where it is stored in a set of secondary or 'accessory genitalia'. This frees the claspers at the tip of the abdomen to grasp the female at the top of her head for the 'tandem' flight (see 125), an activity which precedes and follows mating. During copulation, which lasts 10–15 minutes and takes place on vegetation or rocks, the female bends her abdomen forward to make the 'wheel' formation and receive sperm from the male's accessory genitalia. They then fly off in tandem for male-assisted egg-laying.

Courted by song: many male insects announce their presence to potential mates with a characteristic song

70. (*Facing, top*) A male cricket, *Nisitrus* sp. (Fam.: Gryllidae) (left), serenades a female on Mt. Kinabulu, Borneo. His song is produced by rubbing together specialized areas at the base of his wings. The raised wings act as resonators, although the song of this species is almost inaudible to the human ear. Other species produce a deafening noise, amplified by specially built mud sound boxes, while some are so high-pitched as to be totally beyond the range of the human ear.

71. (*Facing, bottom*) Here, singing has paid off for a male grasshopper, *Chromacris mites* (Fam.: Acrididae), mating in a Peruvian rain forest. Male grasshoppers produce sound by stridulation, the rubbing of small, peg-like projections on the inner face of the hind femur against a specially hardened vein on the forewing. A few species have females which sing, but their stridulatory apparatus is poorly developed. Both sexes have ears, situated on each side of the first abdominal segment.

72. Serenading has paid off, too, for a male bush cricket, *Dichopetala* sp. (Fam.: Tettigoniidae), Mexico. Note that he leans back and is carried passively by the larger female. He thus avoids her massive, blade-like ovipositor. In this ungainly posture, they are liable to predation. Perhaps this is why they make use of the protection afforded by the spines of a prickly-pear cactus (*Opuntia* sp.).

73. At least six different means of sound production are known in shieldbugs (Fam.: Pentatomidae). Here, courtship results in mating in *Edessa nigrispina*, a species from South American rain forests. Although the source of sound is unknown for this species, the commonest method involves the stridulation of bristles on the hind leg against transverse wrinkles or striae on the underside of two abdominal segments.

74. Cicadas are the most accomplished insect singers. Here, a female *Cicada* sp. (Fam.: Cicadidae) (right), France, approached a male while he was being photographed. The song originates in a remarkable 'click' mechanism on each side of the front end of the abdomen. Some species can emit up to 1000 pulses per second. Others are too dangerous for close-up photography — their songs are deafening!

Courtship signalling in flies: gesture, dancing flight, and sound are all part of mate recognition

75. A male stilt-legged fly, *Ptilosphen insignis* (Fam.: Micropezidae), Trinidad, announces his presence and amorous intent to a female with a series of semaphore signals transmitted with his white-tipped front legs. The female responds with her own set of signals. Male micropezids are aggressive and often fight over females, especially when there is a large assemblage of both sexes. The adults are predators of small insects.

76. Mating takes place only when both sexes have completed their full repertoire of signals.

77. (*Facing*) A male hoverfly, *Eristalis arbustorum* (Fam.: Syrphidae), England, in hovering courtship over a female. Male *Eristalis* are magnificent aerial acrobats. They are a familiar sight along hedgerows or the margins of woodland clearings, where they hover at a height of 2–3 m. Their vision is acute and when one spots a female on a flower below, he swoops down to investigate. If it is a female of the right species, he begins courtship, which comprises hovering over the intended mate, while emitting a high-pitched whining buzz with his wings. Sometimes the female flies off to visit other flowers, but the male shadows her so closely that his movements are an exact duplication of the female's. Courtship hovering is resumed each time the female alights on a flower. Just how the female detects the sounds emitted by the male is unknown. Perhaps the feathery parts of the antennae are sensitive to air-borne vibrations like those of the female fruitflies (*Drosophila* spp.). Often, a male will mistakenly court solitary bees, especially *Colletes* and *Andrena* spp., because the female *Eristalis* are such effective mimics.

Courtship and mating in flies: nuptial gifts, pheromones, gesture, and stroking

78. A male assassin fly or dancefly, *Empis livida* (Fam.: Epididae), England, hangs from a leaf while feeding on another fly. He will not complete his meal, but save the prey for presentation as a nuptial gift to a female. Gift presentation by males has evolved several times in insects and is best known in these flies. It is thought that this behaviour placates the predatory urges of the females — in primitive empidids, the males offer no gifts and are often eaten by their mates. It is also possible that females assess the vigour of males by the size and quality of the gift.

79. Mating, and the male supports himself and his mate by hanging from a stem with his front legs, while the female eats a bug presented as a gift. The males of some species congregate in aerial dances, each bearing a gift and females are attracted to the swarm. In others, the males encase a very small gift in a white wrapping of silk, which makes the gift more conspicuous, while advanced species present an empty balloon of silk. The re-use of gifts is widespread.

80. (*Facing, top*) A hitherto unobserved type of courtship. A male fly, *Terellia serratulae* (Fam.: Tephritidae), England, circles a female, rhythmically expanding and contracting a membranous sac which runs the length of the left side of his abdomen. The left side is always inclined towards the female, which suggests that the sac is either a visual signal and/or the source of a mating scent or pheromone. Significantly, the males of this species lack the conspicuous wing patterns typical of tephritids and which are used in courtship semaphore signalling.

81. (*Facing, bottom*) Gesture and stroking are used by males of the robberfly *Cyrtopogon rufipes* (Fam.: Asilidae), France, never before observed courting. A male lands in front of a female, which may or may not have prey. He raises the shiny black tip of his abdomen and bobs it up and down vigorously, while weaving from side to side. As courtship proceeds, he edges closer and closer and strokes the sides of the female's head with his front legs. If the female is receptive, he mounts and copulates with her.

Courtship and mating in butterflies: wing patterns, dancing, aphrodisiacs, and love dust

82. A pair of mating skipper butterflies, *Onophas columbaria* (Fam.: Hesperidae), Trinidad. Mating in butterflies is nearly always the culmination of a more or less complex courtship. Some male skippers have scent-bearing pouches on their forewings, which they flutter in front of potential mates. Visual cues are important in many butterflies and males recognize mates by their species-specific wing patterns. Courtship flights follow and may involve mutual spiralling and/or hovering, the details of which are characteristic of the species.

83. But visual cues can be deceptive, especially when females mimic the appearance of those of other species (see 235–239). Scents therefore come into play once visual contact is established. Here, a pair of *Heliconius isabella* (Nymphalidae: Heliconiinae), Trinidad, reach a crucial stage in their courtship. The male (above) has pursued the female, fluttering above her, and directing a scent towards her from specially modified wing scales. But he is unsuccessful: she signals her refusal by tilting her abdomen upwards.

84. (*Facing*) Scent probably plays a role, too, with this pair of *Euphaedra neophron* (Fam.: Nymphalidae), Kenya. This forest-dwelling species frequents shady areas and feeds on or near the ground. Both sexes have similar wing patterns. The male (above) has a characteristic fluttering flight; several males usually settle near a female when she alights. In many butterfly species, the fluttering disperses the tips of scent-bearing wing scales or androconia, sometimes called 'love dust', which has an aphrodisiac effect on the female. It is not known whether this is the case here. Nor is it known how a female *Euphaedra neophron* selects a mate from her several suitors.

Territories and leks: real estate ownership and display assemblies in male insects

85. Many male insects defend territories, rather like birds and mammals. Here, a male dancefly, *Lissempis nigritarsus* (Fam.: Empididae), England, patrols his minute, 6 square inches of territory. A male with his own territory ensures that only he mates with females which enter. In this and many other species, aggression is ritualized and when males with adjacent territories meet, they veer sharply away from each other.

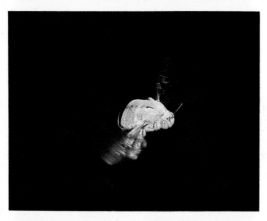

86. Some territorial species are, however, very aggressive. This male carpenter bee, *Xylocopa* sp. (Fam.: Anthophoridae), Kenya, defended his territory around flowers of *Bougainvillea*. He attacked and vigorously butted any male which entered. Male *Xylocopa* have a rapid, darting flight and hover periodically while they inspect their domains. Male insects often set up territories which contain a resource attractive to females, in this case, flowers.

87. Aggressive territoriality is common among the larger hawker dragonflies. This scarlet male *Hadrothemis defecta* (Fam.: Libellulidae) stands sentinel over his territory, a stretch of rocky stream in the Masai-Mara Game Reserve, Kenya. He made short feeding flights and mate-seeking expeditions, but always returned to the one spot from where he had a commanding view of his territory.

88. (*Facing*) Lek behaviour is normally associated with birds and mammals, but more and more instances are now being found in insects. A 'lek' is an assembly of males which congregate at a vantage point and stage a display of some kind which attracts females. Typically, a lek is set up at a place which contains no resource of interest to females, except the prospect of a mate, and the female has a free choice of many competing males which are capable and willing to mate. The males often fight for the best position. Here, a mêlée of male *Lasioglossum calceatum*, a primitively social species of mining bee, leks on the dead flower head of a knapweed. They constantly wheel about with their antennae erect and wings shivering. The wings are semi-irridescent when folded and this, and possibly a scent, attracts the females. These hover around the mass of males, then dart in and are mated. It is not understood just how the female selects her mate. This is the first observation of lekking in any species of European bee.

Mating in mantids: a risky business for the males of some species, but safer for others

89. This mating pair of a species of *Tithrone* (Fam.: Hymenopodidae), Trinidad, remained together for 24 hours, most of it in copulation. The male, as in all mantids, is smaller than the female. Several pairs of *Tithrone* mated in the vicinity and in no instance did the female live up to her ferocious reputation and eat her mate.

90. The small males of this *Acontista* sp. (Fam.: Hymenopididae), Trinidad, also seemed to be immune from attack by their mates. Note that the male is very different in appearance from the female (sexual dimorphism). Here, a second male attempts to oust a successfully mating one. This kind of harassment is common in insects (see 93–95).

91. A male *Polyspilota* sp. (Fam.: Mantidae) tries to mate with a female of a species of *Danuria* (Fam.: Vatidae), Kenya, a genus of stick-mimicking mantids with reduced wings. The male's genital claspers are fully engaged and his mistaken mate carried him around. Although interfamilial matings cannot produce offspring, male *Polyspilota* often indulge in them, possibly because their own females are difficult to catch.

92. (*Facing*) Here, in the Shimba Hills, Kenya, a female praying mantis, *Polyspilota* sp. (Fam.: Mantidae), eats her mate while he continues to copulate with her. A second male of the same species sits hopefully on her back while she completes her grisly meal. The male is able to continue copulation long after his head has been eaten. This is because the process is controlled by a ganglion in the last abdominal segment. Indeed, the removal of the head may well facilitate copulation because this destroys inhibitory centres in the brain. If, however, the male is initially able to grip the female in the right way, he is apparently quite safe. If he does not, the female treats him as an item of prey. Nevertheless, even if he is eaten, the male has perhaps not died in vain. His death can be regarded as a paternal investment in his offspring via the protein of his tissues.

Conduct unbecoming: competition between males. Sometimes competition for mates gets a little out of hand, even spiteful.

93. Troilism in Trinidadian weevils, *Brachyomus octotuberculatus* (Fam.: Curculionidae). A second male tries to mate with an already mating female. Copulating pairs of insects are frequently attractive to males of the same species, which often try to usurp the mating male. In some cases, two males may actually engage their genitalia with a single female. Would-be usurpers may be very violent in their assaults, especially in species where the females mate only once; such males are faced with a rapidly dwindling resource as the proportion of virgin females in a population diminishes.

94. (*Facing, top*) Here, a pair of timber beetles, *Stenocorus meridianus* (Fam.: Cerambycidae), England, has attracted the attention of a wandering male. He apparently detected them by scent, flying around their immediate vicinity, long before settling on their leaf. He immediately leapt on to the back of the mating male and tried to drag him away from the female. First, he seized the male's head in his jaws and tried to pull him off, straining backwards with his legs, but the head provided insufficient purchase. He then gripped one of the other male's antennae and dragged him partly off the female's back. Next, he grabbed the male with his legs and, scrabbling on the leaf margin, shown here, he pulled the male off the female, at an awkward angle, without their genitalia disengaging.

95. (*Facing, bottom*) The interloper rested a short while, then resumed his frenzied onslaught by sitting astride the first male, gripping the edge of the leaf in his front legs, pulling himself over and under the edge, using the leaf's elasticity as a lever. This almost prized the pair apart. The attacking male then suddenly let go, groomed himself, and flew away. The original pair was left permanently, sexually maimed. As shown here, their genitalia were badly injured and could not be disengaged. The male would never mate again and the female could never lay eggs. In this never-before-seen outburst of 'spite', the interloper ensured that if he could not fertilize the female's eggs, then no other male would. We have seen this behaviour in three other species of timber beetle and in no case did the interloper attempt to mate.

4 Investments for the Future
Egg-laying and parental care

When a pair of insects mate, both sexes make an investment on behalf of their genes. The minimum capital is two gametes, a sperm and an egg. The success of the enterprise can be measured in the number of offspring which survive to mate; they are the genetic returns on the investment.

For most male insects, the act of mating is the sole investment. But things are very different for females. An egg is many thousands of times larger than the sperm which will fertilize it. It is therefore costlier to produce and represents a much larger investment on the part of the female. It is not surprising, therefore, that many more females than males provide parental care for their offspring; they have more to lose if they do not.

As shown in the following pages, the simplest form of parental care is to lay eggs in a safe place, close to or in a source of food for the young. This means that the female must be able to make judgements on behalf of her future offspring. And in most cases, this involves assessments of environments and food which are radically different from her own requirements.

Some female insects take parental care one step further and guard the eggs until they have hatched and the young can fend for themselves. The highest forms of parental care are found in dung beetles and solitary bees and wasps. (The special case of social insects is discussed later.) Here, the female constructs a protected space — the nest — and provisions it with all the food required for larval development.

The construction of a nest and foraging for food are costly in terms of time and energy, and nest-building insects therefore lay relatively few eggs. A solitary mining bee, for example, may lay only 15 eggs in her entire life and carpenter bees, *Xylocopa* spp., never lay more than six to eight eggs.

Bees and wasps produce the largest of all insect eggs. Species of *Xylocopa* with a body length of 30 mm lay eggs of up to 15 mm. When one considers this, together with the time and energy spent in constructing and provisioning a brood cell, it is clear that each egg is a massive investment compared with, say, that of a butterfly, which may lay several hundred eggs in one day. This differential in investment is the result of two contrasting strategies: the solitary bee lays few, large, well-protected eggs, a very high proportion of which will survive to become adults; the butterfly, on the other hand, lays a large number of smaller, exposed eggs, relatively few of which will survive the guilds of predators and parasites waiting to assail them and the larvae.

While parental investment is largely a prerogative of female insects, some males do make larger investments than simply copulation. Some male bugs, as we shall see, remain with their females until they lay eggs. They then take charge and guard the eggs. In belostomatid water bugs, the female lays eggs on the back of the male, where they remain until hatching. Both these types of paternal investment could evolve because the eggs are laid immediately after mating: the males are therefore assured of the paternity of the eggs.

Paternity is a problem for the males of species in which the females mate more than once. During copulation, the male deposits sperm in a sac inside the female, called the spermatheca. The female releases a little of his semen every time an egg passes down the vagina, when fertilization occurs. With promiscuous females, a male can never be sure that it is his sperm, and not that of a rival, which fertilizes any given egg. Moreover, it has been shown by genetic studies and sperm irradiation that in many insects, sperm is released from the spermatheca, at least for a short time, on a last in–first out basis. This phenomenon, called sperm precedence, means that the most recent male to mate with a female fathers most of her offspring. Sperm precedence and adaptations to overcome it are recurrent themes in the sex lives of male insects and examples are shown in the next few pages.

Many male insects have evolved bizarre ways to ensure paternity. Some mosquitoes, for example, produce not only a jelly-like 'mating plug', but also a scent in their seminal fluid, termed 'matrone', that renders their mates inviolate and unattractive to other males. And some male butterflies deposit an anti-aphrodisiac scent on their mates, with the same results.

It is well known that the males of some praying mantid species meet a grisly death by being eaten by the females during copulation. Even this can be regarded as a form of paternal investment. First, by remaining in copulation while being eaten, the male ensures his paternity of the female's eggs by excluding rival males. And second, his body, as food, provides some of the raw materials for the eggs, his potential offspring.

Perhaps the most outlandish behaviour is that of male bugs of the genus *Xylocoris*. They try to mate with any females they encounter. They also manage to maximize their reproductive fitness by performing homosexual rape. In so doing, they turn the principle of sperm precedence to their own advantage. When a rape victim next mates with a female, it is the rapist's sperm, rather than his own, which inseminates the female; the rapist has mated by proxy.

No matter how strange and convoluted this behaviour seems to us, it is all geared to one end: producing as many offspring as possible. This is the tyranny of the genes, shaped by the chilling logic of natural selection.

Parental care at its most basic: feeding for egg production and egg-laying in the correct place

96. The eggs within this female dancefly, *Empis trigramma* (Fam.: Empididae), England, are clearly visible through the transparent intersegmental membranes of the abdomen. Many female insects require one or more protein meals in order to bring their eggs to maturity. Here, the dancefly sucks the body fluids of a midge (Diptera: Chironomidae). She will lay her eggs in a damp place such as moss, dead leaves, or rotten wood, where the larvae feed on decaying vegetation.

97. A female alderfly, *Sialis lutaria* (Megaloptera: Sialidae), lays a clutch of several hundred eggs. The larvae live in the muddy bottoms of ponds and streams, preying on worms and larvae. The female lays her eggs near water, to which the larvae migrate. A female parasitic wasp (*Trichogramma* sp.) investigates one of the alderfly eggs, into which she will lay one of her own.

98. Female mosquitoes lay eggs directly in the water in which the larvae will live. Here, a culicid rests on the surface film of a pond and lays several hundred eggs which form a floating mass. The larvae enter the water directly and feed on algae, though some species are carnivorous. They breathe atmospheric air via a respiratory siphon which breaks the surface film of water.

99. (*Facing*) A female King Cracker butterfly, *Hamadryas amphinome* (Fam.: Nymphalidae), lays rows of yellow eggs along the tendrils of a vine in Trinidad. Female butterflies are usually attracted to the larval host plant by chemical cues. The Cabbage White, *Pieris brassicae* (Fam.: Pieridae), for example, is attracted to cabbage plants by the smell of mustard oils in the leaves. Before laying eggs, the females of many species inspect the leaves for signs of eggs laid by other females. If any are present, they move to another, egg-free leaf or plant, thus ensuring that their offspring are free from competition for food. Some tropical vines (*Passiflora* spp.) exploit this behaviour to avoid being eaten: they have evolved 'fake' eggs; raised, yellowish spots on the parts where butterflies would normally lay eggs.

Egg-laying against desiccation: female parental care advances to laying eggs in enclosed, drought-resistant situations

100. A female tiger beetle, *Cicindela campestris* (Fam.: Carabidae), England, uses the tip of its abdomen to scrape a trough in soil into which she lays an egg. Tiger beetles live on dry, sandy soils. The loose soil collapses over the egg and protects it from desiccation. The larva lives in a vertical burrow and preys on insects which stray near the entrance.

101. Female grasshoppers, too, lay their eggs in the ground. Here, *Orthochtha dasycnemis* (Fam.: Acrididae), Kenya, inserts her abdomen into sandy soil. The flexibility of the intersegmental membranes enables her to extend her abdomen to twice its normal length or more, thus making a tunnel long enough to accommodate 30–100 eggs. These are bathed in a secretion which hardens to form a tough pod.

102. A female fruit fly (Fam.: Tephritidae), Mexico, lays eggs via a blade-like ovipositor into a wild fruit. Here, the eggs are deposited directly into the larval food. Several species are major pests of commercially grown fruit. The Mediterranean Fruitfly, *Ceratitis capitata*, is a cosmopolitan pest of citrus fruit.

103. (*Facing, top*) High summer and a female dragonfly, *Aeshna juncea* (Fam.: Aeshnidae), lays eggs in floating rushes in a pond in Gloucestershire, England. The eggs remain dormant, embedded in the plant tissues until the following spring. Sometimes, *A. juncea* lays eggs in vegetation overhanging a dried-up pond, or in plants such as reedmace (*Typha latifolia*), which grow only where seasonal flooding occurs. In either case, water will eventually be available for the aquatic nymphs.

104. (*Facing, bottom*) A female *Aeshna cyanea* (Fam.: Aeshnidae), England, lays eggs in soft, rotten tree roots by a pond. The damp log indicates to the female that winter and spring raising of water levels will inundate the log in time for the hatching of her eggs. While this shows that dragonflies can, *in effect*, assess the potential of oviposition sites, it does not imply 'foresight'. Rather, it shows that insects can be keyed into clues of repeatable environmental events such as flooding.

Mother knows best: many insects lay eggs directly into larval food, or close to it

105. This minute, 3-mm long female weevil, *Balanobius salicivorus* (Fam.: Curculionidae), England, uses the biting jaws at the end of her long rostrum to drill an oviposition hole in a willow leaf (*Salix* sp.).

106. The hole complete, the female turns round and lays an egg in it. When the larva hatches, its presence within the leaf stimulates the plant to form a gall — a kind of tumour, which isolates the larva and on which it feeds (see also 114 and 115).

107. A sawfly, *Tenthredopsis litterata* (Fam.: Tenthredinidae) lays an egg in the stem of an Ox-eye Daisy (*Chrysanthemum leucanthemum*), England. Sawflies are the most primitive of living Hymenoptera; their name refers to the twin saw-like blades of the ovipositor or egg-laying tube, which a female uses to make an incision in plant tissue prior to egg-laying. The blades of the ovipositor move alternately, one being withdrawn while the other thrusts forward, until an appropriately-sized incision is completed. The size of the saw teeth is modified for the substrate used by a particular species. In sawflies which oviposit in wood, the teeth are large and stout, while those which lay in softer, leaf tissue, have much finer serrations. The larvae of sawflies closely resemble the caterpillars of butterflies and moths, which reflects parallel evolution linked to similar feeding habits, not common ancestry.

108. (*Above*) A fly, *Alloeostylus diaphanus* (Fam.: Muscidae), England, laying eggs in the flesh of a Scaly Wood Mushroom (*Agaricus sylvaticus*). The early larval stages (instars) feed on the fungus itself, but the later stages are carnivorous and feed on the larvae of other fungivorous flies.

109. A quintet of bluebottles, *Calliphora* sp., and a lone greenbottle, *Lucilia* sp. (both Fam.: Calliphoridae), lay eggs on the corpse of a young bird which has fallen from its nest. The larvae of both species are important members of the succession of insects which perform the useful function of scavenging on carrion. They predigest the flesh with salivary enzymes and feed on the resulting liquid.

110. This female robberfly (Fam.: Asilidae) uses her spine-like ovipositor to lay eggs in a stick in Oaxaca State, Mexico. Larval asilids are either predators or scavengers and live in wood, soil, or sand, although the habits of the species shown here are unknown.

More egg-laying directly into larval food: wood borers, leaf miners, and gall-makers

111. Preparing an oviposition hole can be laborious. Here, a female timber beetle, *Sternotomis variabilis* (Fam.: Cerambycidae), Kenya, uses its powerful jaws to bit a hole in a dead log prior to laying an egg.

112. When the hole is complete, she turns round, inserts her ovipositor, and lays an egg. The larva will burrow into and feed on the timber for several years.

113. The pattern on this blackberry leaf is the feeding track of a leaf-mining moth caterpillar, *Stigmella aurella* (Fam.: Nepticulidae), England. The female moth usually lays one egg per leaf and the larva tunnels between the upper and lower surfaces. As the larva grows, the mine becomes wider. The mine is characteristic of the species, as is the disposition of larval faeces or frass.

114. (*Facing, top*) Oak Currant Galls, caused by the gall wasp *Neuroterus quercus-baccarum* (Fam.: Cynipidae), England. The wasps alternate between a bisexual generation and an entirely female (agamic) generation, which produces eggs without fertilization (parthenogenesis). The agamic generation appears in April and the females lay eggs deep in the oak catkins and young leaves. This stimulates the development of the galls, within which the wasp larvae feed. They in turn give rise to the bisexual generation.

115. (*Facing, bottom*) The females of the bisexual generation of *Neuroterus quercus-baccarum* lay eggs at the side of veins on the undersides of oak leaves. This stimulates the formation of these familiar spangle galls, which appear in October. The galls eventually drop to the ground with the leaves, and the larvae remain within them throughout winter and in spring give rise to another entirely female, agamic generation. These migrate to the oak catkins and so the cycle of alternating generations proceeds.

Active maternal care: many females remain with their eggs, guarding them for the first few days

116. A female mosquito, *Trichoprosopon* sp. (Fam.: Culicidae), Trinidad, guards her raft of eggs. The eggs are normally laid in water which has collected in old coconut shells. If threatened, the female rows her egg raft into the shelter of the overhanging rim of the nut shell. She remains with her offspring for 2 days after the eggs hatch.

117. A female 'parent bug', the shieldbug *Elasmucha grisea* (Fam.: Pentatomidae), England, guards her nymphs as they feed. She spends 2–3 weeks guarding her diamond-shaped clutch of eggs, entirely covering it with her body. She remains with her nymphs for their first two instars and interposes herself between them and any source of danger; whenever she moves around, the nymphs follow her.

118. Guarding of eggs is common in insects which lay exposed clutches on leaves. Here, a tortoise beetle, *Omaspides* sp. (Fam.: Chrysomelidae), Peru, protects her egg batch from a marauding worker ant.

119. (*Facing*) Here, a female praying mantis, *Tarachodula pantherina* (Fam.: Mantidae), stands guard over her egg pod on a bush in the Meru National Park, Kenya. Although protected from desiccation by the pod, which originates as a viscous secretion of the female, the eggs are still vulnerable to attack by birds, small mammals, and parasitic wasps. The female therefore remains with her egg pod until the nymphs hatch.

Maternal care in a tropical fungus beetle: *Pselaphicus giganteus* (Fam.: Erotylidae), Trinidad

120. (*Facing, top*) A female guards her newly hatched larvae. She is responsible for guiding them to their fungus meal. The female synchronizes egg-laying with fungal development, so she selects a log with fungi just beginning to grow.

121. (*Facing, bottom*) The female *Pselaphicus* leads her brood to a fungus. She continually fusses about them, touching them with her antennae and rounding up stragglers.

122. Two days later, the larvae have already moulted once and are growing rapidly. Here, they begin to feast on a fungus to which their mother has just guided them.

123. Two full-grown larvae take their last meal before pupation. They fed so rapidly and constantly that defecation was a continuous process; the larval stage lasted only 4 days. Parental care of the sort shown here is more likely to evolve in females than males. This is because the female gametes (eggs) are many thousands of times larger than the males' sperms. In other words, any female insect automatically invests more in her offspring than her mate and this alone may commit a female into more investment in terms of guarding offspring, if this significantly increases their chances of survival. Moreover, a female's reproductive success is limited by the number of eggs she can produce, whereas that of a male is limited only by his number of copulations. However, some males do make greater parental investments.

Paternal investment: sperm protection, egg guarding, and male-assisted egg-laying

124. In insects where females mate many times, a male cannot be certain of the paternity of all of the offspring derived from his mates' eggs. His sperm may be diluted or even displaced by that of subsequent males. Post-copulatory guarding of females, while they lay eggs, is therefore common, as seen here in the Yellow Dungfly, *Scathophaga stercoraria* (Fam.: Scathophagidae), England.

125. The same considerations apply in the dragonfly *Sympetrum striolatum* (Fam.: Libellulidae), England, seen here in post-mating tandem flight (see also 69). The male, in front, grasps the back of the female's head with his claspers. They fly over water and periodically bob downwards so that the female's abdomen dips below the surface and some of her 200 or more eggs are released.

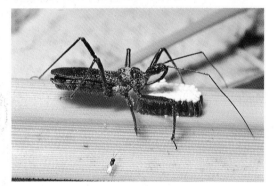

126. This male assassin bug, *Rhinocoris tristis* (Fam.: Reduviidae), Kenya, displays a more direct parental investment. He straddles his mate's egg batch and guards it against female parasitic wasps (Fam.: Trichogrammatidae), one of which is seen here. He chases them away, sometimes impaling them on his rostrum. He does not always succeed, but guarding keeps egg parasitism and predation to an acceptable level.

127. (*Facing*) A male damselfly, *Pyrrhosoma nymphula* (Fam.: Coenagriidae), England, in tandem with his mate while she lays eggs in a submerged rush. Male damselflies clasp the female's prothorax rather than the back of her head, as do most dragonflies (see 125). A male *Pyrrhosoma* is very protective and makes strenuous efforts to rescue his mate if she is caught from below by a water spider or predatory bug. If his mate becomes detached, he turns round and repeatedly bobs in and out of the water, trying to recapture her with his legs. Male-assisted egg-laying ensures not only the paternity of the eggs but also maximizes the chances that the eggs fertilized by a given male are laid in a suitable place. The males of a North American damselfly, *Calopteryx maculata*, ensure the paternity of their mates' eggs by scraping out most of the sperm deposited by previous males, using a scoop-like modification of the 'penis' of the accessory genitalia (see 69). The 'penis' is armed with backwardly directed spines, which rake up the sperm.

Nest building: a protected site for brood rearing is a major advance in parental care

128. A female scarab beetle, *Scarabaeus aeratus* (Fam.: Scarabaeidae), Kenya, rolls a ball of elephant dung to her nest burrow in the ground. There, she lays an egg in the dung and the larva lives surrounded by all the food necessary for complete development. The females of some species plaster the dung ball with a layer of mud, which protects food and larva from desiccation. Females often steal dung balls from each other. Scarab beetles disperse vast quantities of dung.

129. After catching and paralysing a spider, this female spider-hunting wasp, *Batazonellus fuliginosus* (Fam.: Pompilidae), Kenya, excavates a nest in a sandy track. She uses her front legs to dig out sand, which she passes between her hind legs. Wasps and bees are the best exponents of nest building in insects, though pompilids are primitive in this respect — their nests rarely comprise more than an enlarged space or cell at the end of a short tunnel.

130. (*Facing, top*) The female *Batazonellus* drags a spider back to her nest. The spider is alive, but paralysed with venom injected by the wasp's sting. The wasp leaves the spider for a short time near the nest entrance, while she inspects the latter. She returns regularly to the spider, brushing it with her antennae and drives off any ants which may have found it. Inspections completed, she drags the spider into the cell at the end of her nest tunnel and lays a single egg on it. The egg hatches after a few days and the larva feeds non-stop on the spider, moulting several times, until it is fully grown. It then spins a silk cocoon and pupates.

131. (*Facing, bottom*) A vibrating blurr, the female uses her body as a pile driver, tamping down the nest closure of sand scraped back into the burrow. She disguises the nest entrance with loose sand, so that it is completely invisible; she then goes off to search for another spider, with the rapid, agitated gait characteristic of the Pompilidae. Her habit of excavating a nest after prey has been caught is primitive, as is the family-wide habit of providing a single, large item of prey. Other, more advanced pompilids, build nests before seeking prey. Like all solitary (= non-social) wasps and bees, the female *Batazonellus* does not live long enough to see her offspring.

A solitary hunting wasp and its prey: the Common Digger Wasp, *Mellinus arvensis* (Fam.: Mellinidae)

132. (*Facing, top*) A female *Mellinus arvensis* stings her fly prey in the neck. The victim is a Yellow Dungfly, *Scathophaga stercoraria*, itself a voracious predator (see 56), which sometimes overcomes *Mellinus* females seeking prey at dung pats. *Mellinus* shows several behavioural advances over the pompilids (see 129–131): the nest tunnel and first cell are built before prey is caught, several prey items (three to nine) are stored in each cell, and nests have up to 10 cells.

133. (*Facing, bottom*) Here, the female wasp has caught a hoverfly, *Syrphus vitripennis* (Fam.: Syrphidae), stung it in the neck, a method of attack typical of this wasp, and prepares to fly back to the nest with it. The mode of prey carriage is characteristic of the hunting wasp concerned. This species usually flies back to the nest belly-to-belly with the prey, which is always grasped by its mouthparts or antennae in the wasp's jaws.

134. (*Above*) At the nest, the wasp backs into the entrance and drags the hoverfly in after her. When the full complement of paralysed flies is assembled, the wasp lays an egg on one of them, seals the cell and excavates another one at the end of a side branch off the main nest tunnel. *Mellinus arvensis* nests in dense aggregations in sandy soils, often in shaded, overhung banks, and preys on a wide variety of flies.

More solitary hunting wasps: nest building and a miscellany of prey

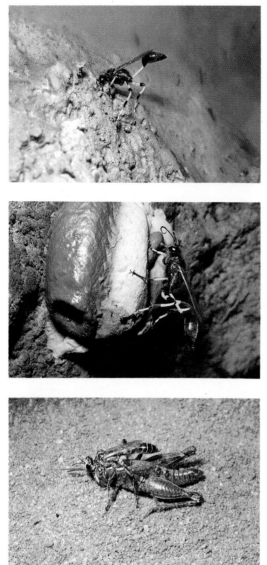

135. Many wasps have independently evolved the habit of constructing exposed nests on rocks, trees, or leaves, using specially collected building materials. Here, a mud-dauber wasp, *Sceliphron spirifex* (Fam.: Sphecidae), Kenya, collects mud from a pond side. Supporting herself on her mid- and hind legs, she moulds a mud pellet with her forelegs and mandibles, then returns with it to her nest.

136. Back at the nest, she uses the mud as the foundation for a cell. The mud of a recently completed cell (left) is still wet and shiny. The cell contains five to twelve paralysed spiders, one bearing the wasp's egg. *Sceliphron* is one of several sphecid genera which have evolved the spider-hunting habit independently of the Pompilidae, and often nests on or in human habitations.

137. Another thread-waisted hunting wasp, *Prionyx kirbyi* (Fam.: Sphecidae), Kenya, drags her prey by its antennae to her single-celled nest in the ground. This species provides just one insect per cell. All *Prionyx* specialize on grasshoppers and, like all the solitary wasps mentioned so far, practises mass provisioning, i.e. each sealed cell is provided with all the food necessary for complete larval development.

138. (*Facing, top*) A species of the highly specialized and worldwide genus *Bembix* (Fam.: Nyssonidae), Kenya, hovering with a paralysed horsefly (Fam.: Tabanidae) over its nest entrance. Unlike the wasps mentioned so far, *Bembix* practises progressive feeding, i.e., she lays an egg on the first fly brought to the nest and then feeds the larva on a daily basis with freshly caught flies. The mother wasp regularly cleans the brood cell of uneaten fly remains.

139. (*Facing, bottom*) Two Peruvian hunting wasps, *Rubrica nasuta* (Fam.: Nyssonidae), fight for possession of a fly. The wasp on the right is the captor and was on her way back to her nest burrow when she was ambushed and knocked to the ground by the other wasp. Prey stealing is common in some hunting wasps. Like their relatives, *Bembix*, the exclusively South American species of *Rubrica* are fast fliers and practise progressive feeding of their larvae.

Hunting and prey manipulation in social wasps and ants

140. (*Facing, top*) The social paper-making wasps are, like their solitary relatives, hunters of insect prey. They are progressive feeders (see 138) like some advanced solitary species, but differ in that they do not present the larvae with whole insects. They process the prey through various stages. First, the worker dismembers the captured insects with surgical precision. Here, a worker *Vespula germanica* (Fam.: Vespidae), England, bites the head off a hoverfly, *Eristalis tenax* (Fam.: Syrphidae).

141. (*Facing, bottom*) The dismemberment continues until only the muscular thorax remains. The wasp returns to the nest with this and passes it to a nurse worker, which chews the remains into a homogeneous lump, a process called malaxation. This is fed to a larva, which exudes a droplet of fluid containing sugars and proteins from its mouthparts, on which the adult feeds. This is called trophallaxis and is an important part of social life (see 155).

142. Social wasps do not restrict themselves to defenceless prey. Here, a worker *Vespula germanica* decapitates the body of a Red Wasp, *V. rufa*, which it has just attacked and stung. Sometimes the captor performs prey malaxation at the site of capture, rather than handing over to a nurse worker back at the nest.

143. A worker ant, *Ectatomma* sp. (Formicidae: Ponerinae), Trinidad, carrying a dead beetle it has found, a species of *Lycus* (Fam.: Lycidae). Primitive ponerine ants scavenge singly on dead invertebrates, though the species of *Ectatomma* often prey on other ants. The warningly coloured *Lycus* beetles are very distasteful to birds, but ants find them palatable. Warning coloration is an evolved response to vertebrate predators.

144. Workers of an African mound-building ant, *Myrmicaria eumenoides* (Formicidae: Myrmicinae), scavenge *en masse* on a dead Giant Snail, *Achatina fulica* in Kenya. Myrmicine ants have stings and sheer force of numbers can enable them to overcome even relatively large animals, though the prey is often weak or dying. Ants are the major predators of invertebrate animals in most parts of the world.

Carpenters and leaf-cutters: foraging and nest building in solitary bees

145. (*Facing, top*) A female carpenter bee, *Xylocopa nigra* (Fam.: Anthophoridae), Kenya, departs from a legume flower. The top of the head and thorax are dusted with yellow pollen, where they brushed against the anthers. Using her front and middle pairs of legs, she will transfer pollen to the scopa on her hind legs, the dense fringe of hairs modified for the transport of compacted pollen. Carpenter bees excavate nest galleries in solid timber and many are temporarily sub-social.

146. (*Facing, bottom*) Female leaf-cutter bees line their brood cells with leaf fragments. Here, *Megachile willoughbiella* (Fam.: Megachilidae), England, cuts an oval piece from the margin of a rose leaf. She uses her powerful jaws like scissors and the whole process takes about 3 seconds. The bee holds the cut piece with her tarsal claws and takes flight just before the piece of leaf is finally detached.

147. Before returning to the nest, a female leaf-cutter bee always alights and adjusts her load. Here, *M. willoughbiella* holds the leaf in her jaws and uses her legs and feet to roll the leaf so that its convex surface is applied to the underside of her abdomen. This is the position all leaf-cutter bees adopt to carry their leaf pieces back to the nest.

148. A completed cell of *Megachile willoughbiella*. This comprises four to six oval leaf pieces gummed together with saliva. The female provisions this with a pasty mixture of pollen and honey, lays a single egg, and then seals the cell with six to ten circular pieces of leaf. There are eight to ten cells per nest, which is usually in a beetle burrow in dead wood.

149. The female *M. willoughbiella* forages at Spear Thistle, *Cirsium vulgare*. The branched body hairs, characteristic of all bees, trap pollen grains, but the abdominal pollen scopa, a series of dense fringes of stiff hairs, seen here, is unique to the Megachilidae. Many are major pollinators of crops; *M. rotundata* pollinates alfalfa in the United States and is provided with artificial nest sites.

5 The Social Option

Social insects are everywhere. The mounds of termites and ants are major features of many landscapes. Ants and termites alone comprise one third of the total biomass of animals in the Amazon Basin; and this calculation includes capybaras, tapirs, and people! In the temperate forests of North America, ants move as much humus as earthworms and they are the dominant middle-level predators in nearly all habitats. The workers of a single colony of the honeybee, *Apis mellifera*, may make a total of 2 to 3 million flower-visits in one day. There is no denying the ecological importance of social insects.

In the following pages, we show many details of sociality in termites, ants, and wasps. Here, we are concerned with the major features of social organization and how it may have evolved.

In ants, wasps, and bees, a colony is a family unit, comprising one or more egg-laying females or queens and a large number of sterile females called workers. Smaller numbers of males are produced, often on a seasonal basis. In highly social (eusocial) species, such as termites, ants, stingless bees, and honeybees, at least two generations coexist in the same nest. There is a clear-cut division of labour: the queen lays eggs, while the workers build nest structures, gather food, and rear young. Work allocation is age-related in many species; a worker passes through several roles — cleaner, nurse, builder, guard, and finally, forager.

In many termites and ants, the worker caste is divisible into two or more size classes, each playing different roles in the economy of the colony. Small workers rear brood, forage, and perform household duties, while the larger ones, called soldiers, are mainly, or exclusively, defenders. Termites differ from other eusocial insects in that the worker caste is derived from males as well as females.

The so-called supremacy of the queen of a social insect colony can be maintained in several ways. In some wasps and bees such as bumblebees, the queen is simply a larger female who is able to exert physical dominance via aggression towards her subordinates. This dominance prevents the workers' ovaries from developing. The queen honeybee secretes a pheromone, 'queen substance', which is passed from worker to worker by direct contact with the queen. This chemical inhibits ovarian development in the workers and prevents them from rearing one of their younger sisters as a new queen. But when the queen's powers wane, and the workers detect less and less queen substance, they alter the diet of some of the larvae and rear them as new queens.

There are several advantages in being social. A populous colony can dominate many resources within the foraging range of its workers. This commanding presence is heightened by the ability of the workers of many species to communicate the whereabouts of a food supply to their nestmates.

In termites and ants, workers may lay a scent trail to guide others to a new supply of food. A worker honeybee informs her sisters by means of the famous 'dance language'. A series of circular runs, the 'round dance', indicates that the food is within 25 m of the nest. When the flowers are further away, she dances a figure-of-eight accompanied by rapid abdominal waggling. The angle from the vertical of the straight run in this waggle dance is the same as the angle between the food source and the sun, as seen from the nest entrance. The dance language is a kind of charade of the recent foraging trip; it was decoded by Karl von Frisch in some of the most elegant behavioural experiments of all time.

Co-ordinated defence is another hallmark of the highly social insects. Communication is at the heart of this, too; it is mediated via alarm pheromones, which are produced in a variety of glands. The alarm pheromone alerts workers to a source of danger.

The defensive tactics of social insects may be suicidal for the individual workers concerned. Workers of an American ant, *Camponotus* sp., literally explode in the face of an enemy. The barbed stings of honeybees and some wasps remain behind in the skin of their victims, still pumping venom and recruiting more workers via alarm pheromones. Meanwhile, the worker dies, disembowelled by the process.

This apparent self-sacrifice raises serious questions about the evolution of social behaviour. How can natural selection favour altruism in a sterile worker caste which cannot perpetuate any genes for altruism? How can natural selection favour a sterile, worker caste?

In 1964, W. D. Hamilton found that the answer lay in the method of sex determination employed by the Hymenoptera, bees, wasps, and ants. Male hymenopterans are derived from unfertilized eggs and therefore have half the normal complement of chromosomes; females are derived from fertilized eggs and have the full complement.

This causes the genes to be shuffled in such a way as to distort the relationships between females. It means that sisters share 75 per cent of their genes, but have only 50 per cent in common with their mother, or their own daughters. Thus, a sterile worker, helping her mother to rear sisters, ensures that more of her genes are propagated than if she had daughters of her own. This is called kin selection and is the most satisfying explanation for the evolution of social behaviour in the Hymenoptera.

A worker hymenopteran is therefore very selfish, far from being a self-sacrificing heroine. Kin selection theory takes the altruism out of altruism.

Primitively social paper wasps: nest building and foraging in
Polistes and *Stelopolybia*

150. A female *Polistes canadensis* (Vespidae: Polistinae), Mexico, collects wood fibres from a dead cactus (*Opuntia* sp.), with which she will make her paper nest. According to species, a nest may be founded by a single female, whose first female offspring behave as workers, or she may be joined by mated auxiliaries. In the latter case, a queen emerges at the apex of a dominance hierarchy. Dominance may be asserted by a variety of ritualized behaviours, depending on the species. In *P. canadensis* it is based on aggression — the most aggressive female becomes the egg-laying queen and the ovaries of the other females degenerate.

151. Other species of *Polistes* assert dominance via differential egg eating. That is, in the race to be queen, females recognize and eat eggs laid by rivals. The female which eats most eggs becomes the major egg-layer and, hence, queen. The ovaries of her rivals regress and they function as workers. Whatever the mode of queen determination, all *Polistes* nests, and those of their relatives, have the same defence against attack by ants. They smear the pedicel from which the nest hangs with a shiny, black, highly repellent abdominal secretion, seen here in a nest of *Polistes dorsalis*, in Mexico.

152. A female *Stelopolybia pallens* (Vespidae: Polybiinae), Trinidad, chews up a caterpillar before returning with it to the nest. This 'malaxation' is exactly the same as that in *Vespula* (see 140–141). Colonies of *Stelopolybia* differ from those of *Polistes* in several ways. There are usually several queens instead of one. Moreover, unlike *Polistes*, where castes are behavioural rather than structural entities, in *Stelopolybia*, queens are usually larger than workers, though there are no clear-cut differences in ovarian size. *Stelopolybia* colonies are also perennial and, like those of *Polybia* (see 160–164), are founded by seasonal swarms.

153. A worker *Polistes versicolor* (*right*), Trinidad, has just shared a bolus of chewed caterpillar with the queen (*left*). The worker will feed a larva with the food, while the queen will eat it herself. Feeding in this way frees the queen from the need to make potentially dangerous hunting trips on her own account. If the queen dies or is killed, the next female in the dominance hierarchy takes over, and her ovaries become greatly enlarged.

154. Another worker from the same colony of *P. versicolor* returns with caterpillar prey and further malaxates it before giving it to a larva. Although most of the bolus will go to a larva, some of it, especially the juices, are eaten by the worker. Analysis of gut contents shows that the workers of social wasps do eat solid particles of prey.

155. The *P. versicolor* worker holds the food bolus just inside the cell entrance while the larva bites off a piece. Before this happens, though, the worker signals her presence by rapidly knocking the rim of the cell with her head, causing a vibration and audible buzz. The larva pops its head clear of the cell entrance and the worker pushes the food against the larva's mouthparts. The larva responds by exuding a droplet of saliva and then tugs at the food held by the worker. When it has detached a piece, the worker goes on to tend other larvae. *Polistes* queens and workers frequently tour the comb, giving the feeding signal without proffering food; they feed on the droplets exuded by the larvae. This trophallaxis provides them with carbohydrates, proteins and, possibly, the enzymes they need but cannot make themselves. The queens apparently require this dietary supplement for egg production.

More primitively social paper-making wasps

156. Females of *Belonogaster juncea* (Vespidae: Polistinae), Kenya, at their nest. *Belonogaster* is found only in Africa and southwestern Arabia. They are the most primitively social of all the paper wasps. Unlike *Polistes*, there is no dominance hierarchy between females and hence no caste differentiation; all females lay eggs and almost all are mated. This species specializes in preying on orb-web spiders.

157. This kind of social organization is called 'quasisocial' and is found also in *Ropalidia*, a species of which from Kenya is shown here. Large colonies may have 500–600 individuals. Nests resemble those of *Polistes* and *Belonogaster* and like those of the latter genus, the adults chew holes in the bottom of brood cells and remove larval faeces. *Ropalidia* is restricted to the Old World tropics.

158. Females of *Stelopolybia pallens* (Vespidae: Polybiinae), Trinidad, pour out onto their nest envelope and adopt a defensive posture: the abdomen is raised and the wings spread in an aggressive gesture in response to their nest being jarred by the photographer. Their behaviour is no idle threat; the barbed stings of polybiines, like those of the honeybee, remain in the skin of an attacker.

159. (*Facing*) The vertical comb of *Polistes instabilis* (Vespidae: Polistinae), suspended in a tree in semi-desert country, Tamaulipas State, Mexico. Eggs and larvae are clearly visible in some of the cells. The workers often store nectar and fruit juices temporarily in peripheral cells containing eggs; the food is usually used up before the eggs hatch. The white caps closing seven of the lower cells are silken cocoons spun by the mature larvae prior to pupation. Some of the open cells containing eggs and larvae have tattered and frayed rims. This indicates that the cells have been used for brood rearing on previous occasions, the rough edges resulting from emerging adult wasps biting away the cell caps. In many *Polistes* species, only the queen and co-foundresses initiate new cells, but all females work on their enlargement. The vertical comb of *P. instabilis* is unusual for the genus. Most species build horizontal combs. Note the short but very broad pedicel, with its shiny, black, ant-repellent coating secreted by the wasps.

Growth and development of a colony: the Central and South American paper wasp, *Polybia occidentalis*

160. A swarm has established a new nest beneath a leaf in Trinidad. Colonies of *P. occidentalis* (Vespidae: Polistinae) are perennial and persist for between 4 and 25 years, with regular cycles of swarming, usually after each new generation of males and queens has emerged. A swarm comprises a group of workers and one or more mated queens. Colony fission by swarming, sometimes called 'sociotomy', is normally associated with the four species of honeybee (*Apis* spp.), but is apparently widespread in the polybiine wasps.

161. A few days later and the nest begins to assume its typical globular shape. Meanwhile, some workers will have started to forage for food. *P. occidentalis* stores considerable amounts of nectar, which is used by humans in Mexico as a source of honey. The species preys extensively on winged termites during their nuptial flights and large numbers of biting flies, collected from around the eyes of cattle, are another source of protein for the growing wasp larvae. Nests of *P. occidentalis* contain between 130 and 230 workers and a very few large colonies may contain nearly 600 individuals. There are 4 to 48 queens, which are unusual in being, on average, smaller than the workers but structurally identical to them, except for the greater development of the ovaries and consequent distension of the abdomen. *P. occidentalis* is very widespread, ranging from southern Texas, through Central and much of South America, to Paraguay.

162. A group of workers adds a new layer of cells, which appears here as an irregular, equatorial line of clearly damp and newly made paper. Like all paper-making wasps, *P. occidentalis* uses a pulp of chewed wood or vegetable fibres as a building material. One species, however, *P. emaciata*, from Central and South America, makes delicate and beautiful cells of mud.

163. Construction continues with an outer coat or envelope of carton material. This eventually covers all the horizontal combs, leaving one small entrance hole. Apart from providing mechanical protection, the envelope helps to maintain a fairly constant environment within the nest. The wasps control temperature and humidity in ways reminiscent of the honeybees. Workers may simply stand at the nest entrance and fan cool air inwards, using their wings to create an air stream. They may also bring droplets of water inside and use the air stream to evaporate the water, which cools the nest. *P. occidentalis* overlaps with another species, *P. scutellaris*, in some parts of its range. The two species are virtually indistinguishable, but their nests are very different. That of *P. scutellaris* has spiny outgrowths from the envelope, which may deter predators seeking honey and grubs. The Guarani Indians of Paraguay distinguish between wasps with smooth nests, 'kavichui' and those with spiny nests, 'kamuati'.

164. Alarmed by a sudden jarring of the nest, an angry horde of workers pours out, prepared to reject any intruder. They are armed with painful stings, which, like those of the honeybee, are barbed and remain in the victim's skin. Bats sometimes bite through the envelope to eat honey and grubs, but army ants (see 169–170), are the wasps' worst enemy.

The nests of termites and ants: the insect metropolis

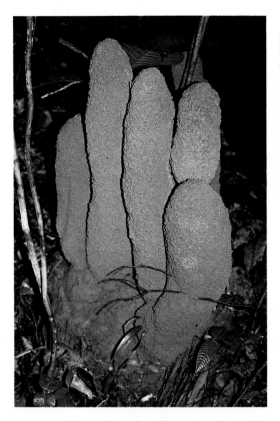

165. The finger-like nest mounds of a ground-nesting termite, *Dicuspiditermes* sp. (Fam.: Termitidae), point skywards in deep forest in Borneo. The mounds are made of earth mixed with faecal pellets comprising partially digested cellulose and lignin, an indigestible component of wood. All termites are highly eusocial, i.e., there are well-developed castes. Although often called 'white ants', termites are not true ants; they form an order of their own, the Isoptera. Despite their apparently ant-like ways, termites might well be regarded as social cockroaches. Primitive termites share many characteristics with the cockroaches and there are similarities in the symbiotic organisms which live in their guts and on which both depend. The symbionts enable termites to digest the cellulose products of plant tissues and include Protozoa (single-celled animals), fungi, and bacteria. *Dicuspiditermes*, like all the higher termites, has only spirochaete-like bacteria in its gut.

166. Some true ants have highly specialized relationships with plants. Here, workers of *Crematogaster nigriceps* (Formicidae: Myrmicinae) defend their nest in the swollen thorn base of a 'whistling thorn', *Acacia drepanolobium*, in Kenya. With their abdomens uptilted, dispensing venom via their stings, they emit an alarm pheromone from the mandibular glands, which recruits other workers. They excavate nest cavities in the pith of swollen thorn bases. The hard skin of the swellings affords the ants and their brood mechanical protection and cushions them from extremes of temperature. Apart from nest sites, the tree provides the ants with two sources of food: nectar, from nectaries among the leaves, and oils and proteins in little outgrowths on the anthers, which the ants harvest and take to the nest. The ants' contribution to this remarkable symbiosis is to keep in check any harmful herbivorous insects and to deter grazing mammals with their very painful bites.

167. Workers of the Green Weaver Ant, *Oecophylla smaragdina* (Formicidae: Formicinae), Australia, fold down a leaf margin to form the walls of a nest; weaver ants build nests out of living leaves. In a remarkable feat of co-operation, the larger workers hold down the leaf edge while smaller workers weave the leaves together in a side-to-side motion of the head, using silk extruded by larvae held in their jaws.

168. A clay nest mound of *Odontotermes latericius* (Termitidae: Macrotermitinae) in Kenya, with vertical chimneys which ventilate the nest interior. *Odontotermes* is restricted to the Old World tropics and cultivates a specialized fungus, *Termitomyces* sp., on which the termites feed. The fungus grows on a network of combs comprising termite faeces and partially digested wood. The termites eat fungus and comb alike. *Odontotermes* often destroys the structural timbers of houses.

169. Army ants, *Eciton burchelli* (Formicidae: Dorylinae), Trinidad, return to their temporary bivouac with a centipede which they have overpowered and killed. Army ant colonies stage swarm raids: a colony of between 150 000 and 700 000 individuals may send out a column perhaps 105 m long and 8 m wide, which fans out on a broad front, flushing out and killing a wide range of leaf-litter invertebrates.

170. Workers of *Eciton burchelli* form a living bridge en route to a new bivouac site. *Eciton* has no permanent nest and alternates between cycles of daily changes of bivouac and temporary stays of up to 3 weeks. The worker caste comprises small- and medium-sized ants and large, defensive soldiers with long, curved jaws. All are blind, and raiding swarms follow scent trails laid by scout workers.

171. Driver ants, *Dorylus nigricans* (Formicidae: Dorylinae), attack a shelled slug in Kenya. The slug's frothy, defensive secretion entraps the ants. Large-jawed soldiers, however, retrieve their helpless worker colleagues and the raiding column retreats. The majority of victims nevertheless succumb to the efficient hunting tactics of *Dorylus*, which occupies the same niche in the Old World tropics that *Eciton* occupies in the New World. The African driver ants have broad-fronted raiding swarms at the head of a marching column, much as in *Eciton*. They are major predators of termites and even cope successfully with lizards and snakes. *Dorylus* will kill and dismember large animals such as domestic fowls, pigs, and monkeys if these are confined by man and cannot escape or if they are immobilized by injury. The bivouacs of *Dorylus* species are more stable than those of *Eciton* and are often excavated deeply in soil.

172. Workers of a carpenter ant, *Camponotus* sp. (Formicidae: Formicinae), feed on the juices of a cactus (*Opuntia* sp.) in the Mexican desert. This is a valuable source of moisture for many desert ants. The workers of *Camponotus* secrete a pheromone from abdominal glands, which recruits other workers to join those which have found a food source. *Camponotus* is a cosmopolitan genus. The species are sometimes called carpenter ants because they usually excavate their nest galleries in trees and their root systems, though some nest in the soil.

173. A leaf-cutter ant, *Atta* sp. (Formicidae: Formicinae), Trinidad, with a piece she has just cut out of the margin of a leaf. In their vast underground nests the ants use chewed leaf fragments as a substrate for the cultivation of a specialized fungus on which they feed. Leaf-cutter ants are the dominant herbivorous species in Central and South America.

174. Workers of a fungus-growing termite, *Macrotermes* sp. (Termitidae: Macrotermitinae), Malaysia, harvest newly fallen dead leaves. When digested and converted into faeces, the leaves will be the substrate for fungus cultivation, on a comb made of finely chewed wood (see also 168). Fungus-growing termites are responsible for returning important nutrients to the soil, though in Malaysia some species are also horticultural pests.

Termites and ants: defence of foraging workers by a soldier caste

175. An entirely defensive soldier sub-caste has evolved many times in both the ants and termites. Here, a soldier of the African driver ant, *Dorylus nigricans* (Formicidae: Dorylinae), guards a column of foraging workers. It shows the adaptations typical of soldier castes — large body-size and a massively developed armoured head, containing powerful muscles for the huge, sickle-shaped jaws. The soldiers of *Dorylus* adopt this defensive stance facing outwards along the margins of columns of the much smaller workers.

176. Soldier termites, too, may have enlarged heads with powerful jaws. Here, a soldier of a *Macrotermes* sp. (Termitidae: Macrotermitinae) guards workers as they harvest dead leaves. But mechanical defence is not the only weapon in the soldiers' armoury. Most species also have some means of chemical warfare at their disposal. Termites such as *Macrotermes* produce a brown, corrosive fluid from the mandibular glands, which is highly repellent to ants which are their main enemies. It is fatally toxic to them in high doses.

177. (*Facing, top*) There are many parallels in behaviour between termites and ants. Here, large-headed soldiers of the same Malaysian species of *Macrotermes* line up on either side of a column of workers harvesting leaves, rather like the soldiers of the driver ant, *Dorylus nigricans* (see 175). In *Macrotermes*, the soldiers are all derived from female nymphs, but in other genera, soldiers comprise non-reproductive individuals of both sexes, while in some (see 178) they are all male. The soldiers of genera such as *Pericopritermes* have asymmetrical mandibles which operate against one another with a snapping action, rather like the sudden snap between human thumb and fingers. This enables the soldier to flick away marauding worker ants.

178. (*Facing, bottom*) Chemical warfare reaches it apogee in nasute soldiers of termites such as this *Nasutitermes* sp. (Termitidae: Nasutitermitinae), Trinidad. The soldiers have no jaws and the head is elongated into a prominent 'nasus' or median nose. This functions like a gun barrel for the forcible expulsion of a defensive chemical secreted by a gland in the head. Despite their blindness, the aim is accurate and the noxious stream has a range of several centimetres. The fluid may be poisonous to ants and or sticky, and inactivates other insects by gumming them up. After discharging its weapon, a soldier wipes its nasus on the ground and returns to the nest, resuming active service when more fluid has been secreted.

II Designed for Survival
Insects and the outside world

6 Parasites, Nest Scavengers, and Cuckoos

We are used to the idea of insects as parasites. Bugs, mosquitoes, and horseflies pay temporary visits to their vertebrate victims, including man, just long enough for a meal; they may attack a wide range of species. Lice and fleas, on the other hand, are highly adapted to their hosts' physiology and are often restricted to one species; they cannot live away from their hosts for long. Ectoparasites rarely kill their hosts, though they may transmit disease organisms.

About 10 per cent of all insects, however, are parasites of other insects. Some, like adults of tiny biting midges of the genus *Forcipomyia*, suck the blood of large insects, such as dragonflies and moths. They do not kill their hosts and therefore resemble the ectoparasites of vertebrates. But the majority of parasitic insects have free-living adults and it is the larvae which are parasites in or on the bodies of other insects, usually larvae. They are called protelean parasites. Because they eventually kill their hosts, they are not typical parasites, but, rather specialized predators. The term 'parasitoid' has been coined for them.

Flies, beetles, or even moths may be parasitoids but by far the largest number are Hymenoptera. At least 50 per cent of the 200 000 known species of Hymenoptera are parasitoids and most of these are ichneumon wasps. Many aspects of hymenopterous parasitoids are illustrated in the next few pages, including adaptations for finding hosts and laying eggs in or on them.

Many parasitoids are host-specific or are at least restricted to a particular group of related hosts. This is valuable for agriculture because host-specific parasitoids can be used as biological control agents for crop pests. While this is an attractive concept, not least because it by-passes the use of pesticides, it has achieved only limited success. Of the 91 pest species for which biological control has been attempted in North America, it has only worked well for 18. The failures were often due to an incomplete understanding of the dynamics of host and parasitoid populations.

The large amounts of faecal material and corpses produced by colonies of social insects are rich potential food sources. It is not surprising, therefore, that many insects are highly adapted for life as nest scavengers. Many species placate aggressive workers with a sugary secretion. Others minimize contact with workers by confining themselves to the lower regions of nests, where waste middens accumulate. Some species which scavenge in the nests of ants have evolved a remarkable resemblance to the shape of the host workers. This, combined with the absorbed nest odour, confers protection against the predacious ants.

Many species have free-living adults and only the larvae scavenge in social insect colonies. The females of such insects may have to run the gauntlet of aggressive workers when they enter the nest to lay eggs. Females of a fly,

Pholeomyia texensis, avoid this in a very neat way. Their larvae scavenge in the nests of the Central and South American leaf-cutter ants, *Atta* spp. The female gains access to the nest by attaching itself to a leaf fragment being carried by a returning forager. The worker ant has its jaws full and could not attack the fly, even if it noticed the stow-away. Females of an African fly, *Stomoxys ochrosoma*, get the ants to carry their eggs into the nest for them. Just how they do this is shown in the remarkable photograph 179.

Among bees and wasps there are many which have become cuckoos: they lay their eggs in the nests of other species. The larvae eat the food stored by the host female. Several of their adaptations for this life-style are described in the following pages.

It has recently been shown that the males of the cuckoo-bee genus *Nomada* have evolved an unusual way to increase their reproductive success. They seek out and patrol the nest sites of the hosts, species of *Andrena*, a large genus of solitary mining bees. The males emit a scent which is the same as that used by female *Andrena* to mark their nest entrances. This increases the attractiveness of the nest site to female *Nomada* and helps individuals which have not yet found a host population to find their way. This odour mimicry not only attracts mates for the male *Nomada*; it also ensures that a female cuckoo, once mated, has to search no further for suitable host nests in which to lay eggs.

The ovaries of cuckoo wasps and bees are also highly adapted for their way of life. A female cuckoo produces two to three times more eggs than its host. She invests no time in nest construction and provisioning, and can therefore devote all her resources to eggs. Her ovaries also have more mature eggs at any given time than those of her host. This enables her to seize the chance of laying as many eggs as possible once a host population has been found.

Parasitoids, nest scavengers, and cuckoos are finely tuned to the opportunities provided by their hosts. They are the result of the exquisite precision of natural selection.

A life of stealth: nest scavengers or parasites, many insects have highly specialized lifestyles

179. *(Facing)* A female fly, *Stomoxys ochrosoma* (Fam.: Muscidae), Kenya, releases a clutch of about 20 eggs while hovering motionless over a column of army ants, *Dorylus nigricans*; she chooses a worker without booty and releases her eggs with great accuracy, placing them within reach of the ant's antennae and jaws. The ant takes them to the bivouac, where the hatched larvae live as scavengers. The fly's egg-laying tactics exploit the drive of worker ants to carry brood, at the same time providing her offspring with a place to live and feed, without herself running the gauntlet of a host of voracious predators. This is the first time egg-laying in *Stomoxys ochrosoma* has been photographed. Hitherto, it was thought that the female laid a single large egg, or gave birth directly to a newly hatched larva.

180. The larva of a spider-wasp (Fam.: Pompilidae) eats an Australian spider. The larvae of most pompilids feed on a permanently paralysed spider in a nest prepared by the mother wasp (see 129–131). However, the larvae of a few species live as external parasitoids on a spider which remains paralysed for only a short time. The spider resumes mobility and feeds normally, but, if female, lays no eggs. It eventually dies and the wasp larva spins a cocoon and pupates.

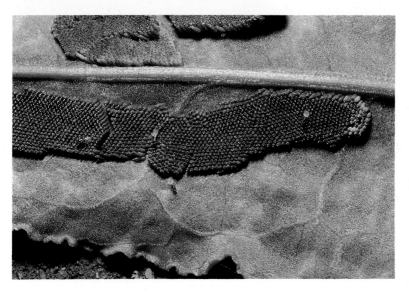

181. A quartet of minute female parasitic wasps, *Trichogramma semblidis* (Fam.: Trichogrammatidae), lays eggs in those of the alderfly, *Sialis lutaria*. Trichogrammatid wasps spend their entire larval lives within the eggs of other insects. *T. semblidis* also develops in the eggs of moths. Up to 90 per cent of host eggs may be parasitized and for this reason, *T. semblidis* was used as a biological control agent for codling moth as early as 1912.

Egglaying on a hidden host: parasitic wasps use an elongate ovipositor or egg-laying tube

182. A female chalcid wasp, *Torymus chloromerus* (Fam.: Torymidae), about to lay eggs on a concealed larva of the Knapweed Gallfly, *Urophora jaceana* (Fam.: Tephritidae), England. The host larva induces the formation of a gall in the flower heads of Brown Knapweed, *Centaurea jacea*; the wasp uses her long ovipositor, seen here, to penetrate the wall of the gall. She may lay eggs singly or in groups. In the latter case, the first larva to hatch eats the other eggs before consuming the host. The wasp lays eggs in August and development is rapid; some adults appear in September, while other individuals overwinter as larvae. *Torymus chloromerus* is one of five wasp parasites of the gallfly. Sometimes, two or more species of chalcids, in addition to *Torymus*, may lay eggs in a host and many *Torymus* larvae die because of this 'superparasitism'.

183. (*Facing, top*) Many parasitic wasps can penetrate hard, woody structures with their ovipositors. Here, a female braconid wasp, *Iphiaulax* sp. (Fam.: Braconidae), oviposits in a woody stem gall in Kenya. The host of this species is unknown, but other species of *Iphiaulax* parasitize the caterpillars of wood-boring moths or the larvae of timber beetles. It is possible that the female wasp detects the host larva by vibrations, but in other species, it is known that scent plays a part (see 186). The female probes the woody surface with her ovipositor until she finds a small fissure, which she then penetrates. The ovipositor drills through the wood by means of the blade-like tips of its component valves, which have minute serrations and move up and down relative to one another.

184. (*Facing, bottom*) The larvae of gall-forming wasps are particularly prone to parasitism of one sort or another. Here, a chalcid wasp, *Eurytoma brunniventris* (Fam.: Eurytomidae), oviposits in a 'pea gall' of oak, formed by the larvae of the wasp *Cynips divisa* (Fam.: Cynipidae), England. The *Eurytoma* larvae feed on the gall tissue and also eat the *Cynips* larvae, although the latter form only a small part of their diet. Strictly speaking, therefore, *Eurytoma* is a 'cleptoparasite', for it feeds largely on the host's food. *E. brunniventris* has been reared from the galls of 49 species of cynipid wasp. Within the galls of *Cynips divisa*, it is only one of 12 species which parasitize the gall maker, and its larvae will also eat these. *Eurytoma brunniventris* is one of a complex of closely related 'sibling' species, in which the adults are virtually identical; they can be distinguished on the basis of egg structure and host relationships.

Parasites and hyperparasites: the Alder Woodwasp, *Xiphydria camelus* and its enemies

185. A female Alder Woodwasp, *Xiphydria camelus* (Fam.: Siricidae), England, uses her ovipositor to drill into a dead alder. As well as introducing one or more eggs, she deposits spores of a symbiotic fungus, which are stored in a pair of sacs opening into her oviduct. The fungus grows in the galleries excavated by the wasp larva and the larva feeds on the fungus as well as the wood for about 10 months. The larva deposits dense masses of faeces or frass in the tunnels behind it. The mature larva pupates just inside the bark and at right angles to it.

186. Large woodwasp larvae are a potentially rich source of food for any insect which can penetrate wood, and two species of ichneumonid do precisely this. Here, a female *Rhysella approximator* (Fam.: Ichneumonidae) searches an alder log for signs of the host. It used to be thought that the females detected the larvae by vibrations caused by its feeding movements. However, it is now known that the ichneumon is attracted by the smell of larval frass contaminated with symbiotic fungi.

187. Having found an Alder Woodwasp larva, a female *Rhysella approximator* drills into the wood with her ovipositor. She may penetrate 2 cm; the ovipositor is supported between the bases of the hind legs. Sense organs at the tip of the ovipositor allow her to assess the suitability of the host larva. If it is fully grown, or nearly so, she lays an egg on it. The egg has a thin, elastic-walled stalk. The substance of the egg is extruded down the stalk, in the ovipositor, so as to reduce its diameter. It resumes its normal shape on emergence at the tip.

188. Her egg laid, the female *Rhysella* removes her ovipositor from the wood and grooms it with her hind legs. Female ichneumonids, and other parasitic wasps which lay eggs on or in deeply concealed hosts, have very long ovipositors, often several times the total body length. Such ovipositors clearly cannot be retracted into the body and are therefore permanently visible. The larva of *Rh. approximator* feeds rapidly on the woodwasp larva and completes its development in 16 days. It then remains dormant for 10 months before entering a pupal stage of 14 days. The adult wasp eventually bites its way to the outside.

189. A female of *Rhysella approximator* has been spotted by that of another ichneumon, *Pseudorhyssa alpestris* (Fam.: Ichneumonidae). This cleptoparasite is unable to drill an oviposition shaft for herself. Instead, she uses one made by *Rh. approximator*. Although *Pseudorhyssa* females often find oviposition holes by direct observation of *Rhysella* females, they can find them in the absence of the other wasps: they are attracted to vaginal secretions left by *Rhysella*. The *Pseudorhyssa* larva kills that of *Rhysella* and it eats the woodwasp larva.

190. Males of *Rhysella approximator* wait for mates at their emergence sites. These two are mistakenly attracted to the rasping made by an emerging Alder Woodwasp; one of them is sufficiently aroused to attempt mating with the woodwasp's head. The Alder Woodwasp has to contend with more than just *Rhysella* and its cleptoparasite *Pseudorhyssa*. Its eggs are parasitized by the larvae of another wasp, *Aulacus striatus* (Fam.: Evaniidae), which lays its eggs down the channel drilled by the woodwasp. Also, the minute chalcid, *Xiphydriophagus meyerinckii* (Fam.: Pteromalidae) lays up to 30 eggs on a single woodwasp larva after paralysing it with venom injected via her ovipositor.

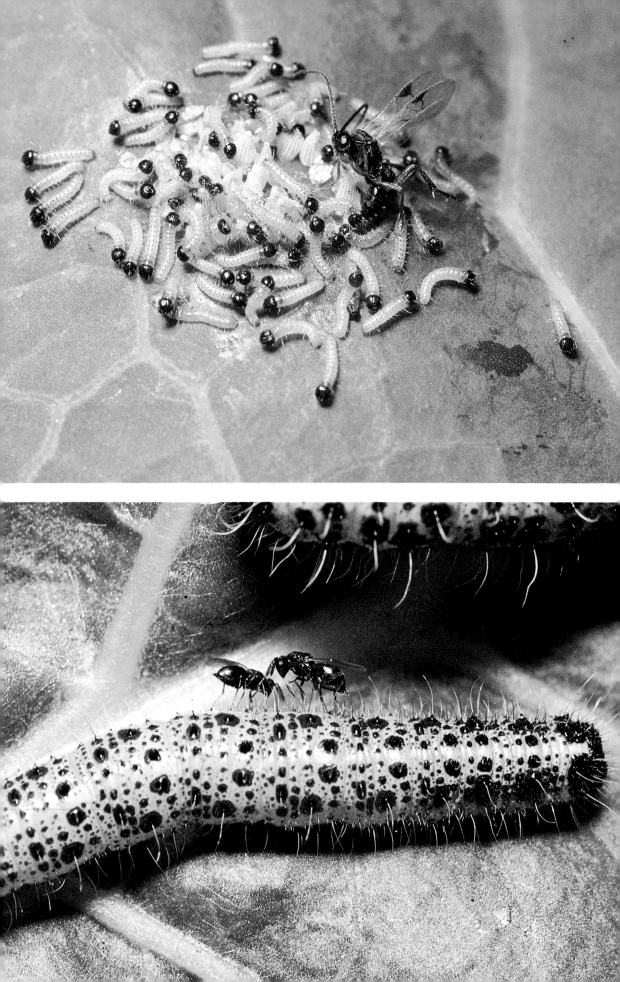

Insect enemies in the service of man: parasites of the Cabbage White Butterfly, *Pieris brassicae*

191. (*Facing, top*) Cabbage growers have several allies in their annual battle against caterpillars of the Cabbage White Butterfly, *Pieris brassicae* (Fam.: Pieridae). Here, a female parasitic wasp, *Apanteles glomeratus* (Fam.: Braconidae), lays eggs in newly-hatched, first-instar caterpillars. The host continues to feed and moult its skin and is finally killed during the last instar. A single caterpillar may support as many as 50 *Apanteles* larvae.

192. (*Facing, bottom*) Older caterpillars attract minute chalcid wasps, *Tetrastichus galactopus* (Fam.: Eulophidae). Here, one of two females oviposits into a Cabbage White caterpillar already parasitized by *Apanteles* larvae. The chalcid lays her eggs directly into the body of an *Apanteles* and her offspring therefore develop as 'hyperparasites'. The females of *Tetrastichus galactopus* can detect the presence of *Apanteles* larvae without probing with their ovipositors. Just how they do this is unknown, but perhaps parasitized caterpillars have a characteristic smell.

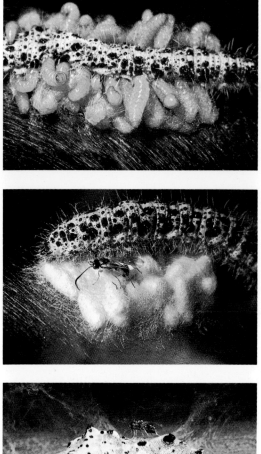

193. A final-instar Cabbage White caterpillar dies, reduced to a husk by about 40 full-grown larvae of *Apanteles glomeratus*, which bite their way out through its skin. Each *Apanteles* larva spins a tough cocoon as soon as it is free. Soon, the remains of the caterpillar will be enmeshed in a mass of flocculent silk as the parasite larvae pupate.

194. The *Apanteles* larvae may escape attack by the hyperparasite *Tetrastichus* (see 192), only to fall victim, as pupae, to another wasp, *Lycibia nana* (Fam.: Braconidae). Here, the female inserts her ovipositor through the *Apanteles* cocoon and lays an egg on the pupa inside.

195. Meanwhile, the Cabbage White pupa (chrysalis) risks attack by yet another wasp. Here, a female chalcid, *Pteromalus puparum* (Fam.: Pteromalidae), explores the pupa before laying eggs in it. The chalcid can penetrate only the soft skin of newly-formed pupae; it rides on the backs of late-instar caterpillars to ensure its presence at the time of pupation and, therefore, a host with a soft skin.

Cuckoos and parasites: enemies of nest-making bees and wasps

196. A cuckoo bee, *Thyreus* sp. (Fam.: Anthophoridae), Kenya, feeding at a flower. As the name implies, cuckoo bees are social parasites in the nests of other bees. A female cuckoo lays her eggs in the brood cells of another bee while the mother is away, foraging. The cuckoo's egg hatches before that of the host. The first-instar larva of the cuckoo has large, sickle-shaped mandibles, with which it destroys the host egg or young larva. It then moults its skin and, replacing its giant jaws with a normal sized pair, feeds on the host's pollen store. The cuckoo habit has evolved independently in four families of bees; the females have lost the nest-building habit and lack the pollen-collecting apparatus of normal bees. Cuckoo bees are usually host specific and some are closely related to their victims. The hosts of *Thyreus* spp. are solitary anthophorid mining bees, usually *Amegilla* spp.

197. (*Facing, top*) A velvet 'ant', *Hoplomutilla opima* (Fam.: Mutillidae), Trinidad. Despite the name, mutillids are not true ants, but a family of wasps whose larvae are solitary ectoparasites of the larvae or pupae of other insects. Relatively few host records are known for this world-wide family. Most species apparently parasitize other wasps or bees, though some are hyperparasites of beetle larvae and several species attack the pupae of tsetse flies (*Glossina* spp.) Female mutillids are always wingless and the thoracic segments are more or less fused. The sexes are strikingly different; the males of many genera have independently evolved modifications of the jaws and head for transporting the wingless females in flight. Female mutillids have an extremely painful sting and one species is reputedly fatal to man; most species have conspicuous warning coloration (see also 231–234).

198. (*Facing, bottom*) The brilliant metallic colours of this ruby-tailed wasp, *Chrysis ignita* (Fam.: Chrysididae), England, are characteristic of this cosmopolitan family. The larvae of most species are solitary ectoparasites of full-grown wasps or bee larvae. *Ch. ignita* lays its eggs in the cells of solitary mason wasps, *Odynerus* and *Ancistrocerus* spp. (Fam.: Eumenidae). The female *Chrysis* lays an egg while the host female is away from her nest. The *Chrysis* larva delays its development until the host larva has eaten all its food. It then hatches and feeds rapidly on the host for 6 days, finally pupating within a silken cocoon. Other species have a typical 'cuckoo' life-style and eat the host's nest provisions after destroying the egg or young larva. Female chrysidids have no sting; instead, all but three of the abdominal segments are fused into an egg-laying tube.

7 Defence in a Hostile World

Insects are surrounded by enemies. They have evolved an enormous array of defences, many of which are described and illustrated in the following pages. Almost all these defences have evolved in response to predators, especially birds.

But how can natural selection have brought about the near-perfect camouflage of many insects, the mimicry of distasteful species by palatable ones, and the uncanny resemblance to inedible objects? How, indeed, can natural selection be invoked to explain the many beautiful adaptations which occupy much of this book?

The principle of natural selection is simple. All plant and animal species are variable and much of this variation is genetically controlled. Any variant which renders an animal more efficient at utilizing a resource or escaping predators enables that animal to produce more offspring than less efficient individuals. In this way, the genes which confer the benefit increase in frequency in the population over time. We say they have been selected for. By the same token, genes for less beneficial variants decrease in frequency; they are selected against.

Thus, adaptive change is all about differentials in reproductive success. As Gore Vidal wrote, in another context, 'It is not enough to succeed. Others must fail.' (*Messiah*, 1955).

The processes of natural selection have now been amply and conclusively demonstrated in field and laboratory experiments. But one of the most potent experimental proofs began less than 200 years ago. It took place not in the laboratory, but in the increasingly polluted countryside of Victorian England; and the experimenter was not a white-coated scientist, but, inadvertently, the Industrial Revolution.

The Peppered Moth, *Biston betularia* is a common species. It roosts by day on tree trunks, with its wings spread out. The normal colour pattern is a fine speckling of black spots and bands on a white background. It is well camouflaged: tree trunks in unpolluted areas tend to be covered with rough-textured lichens. Against this background, the Peppered Moth is difficult to see and predation by birds is slight.

In the 1850s, a black or melanic form of the moth began to appear. It was aptly named *carbonaria* and increased in frequency until 80–90 per cent of some populations consisted of the melanic form. Now lichens are very sensitive to airborne pollution and disappeared from trees in industrial areas and countryside downwind of them. The tree trunks became black, and against this background the normal form of the Peppered Moth was very conspicuous and easy prey to birds. In the jargon of evolutionary biology, birds were the *selective agents* and the normal form of the moth was at a *selective disadvantage* in polluted areas. Prior to the Industrial Revolution, it was the melanic form which was at a selective disadvantage.

The black colour form of *carbonaria* is controlled by a single gene. Natural selection, in the form of bird predation, increased the frequency of this gene in populations in polluted areas. The birds, of course, did not consciously select any particular colour pattern; they simply caught those moths which were easiest to see. With smoke abatement policies having taken effect, lichens are reappearing on tree trunks and the frequency of normal, pale coloured Peppered Moths is rising again.

Our example of natural selection in action is a very simple one; it involves selection acting, via a single agent, birds, on one facet of an insect's life-style. The story of the Peppered Moth is a classic simply because it was conspicuous to human observers and took place rapidly in very recent history.

The struggle for survival, however, is fought on a broad front; selection operates via many different environmental agents and influences not only appearance, but all the complexities of behaviour and the usually invisible processes of development and physiology.

Most of the defence strategies shown in this book are much more complex than that of the Peppered Moth. They required intense selection pressure, acting over hundreds of millions of years. Miriam Rothschild has described the formidable powers of just one insectivorous bird. Her pet Shama, *Kittacincla malabarica*, could spot a flea after it had jumped and then leap after it before it landed. This is no mean feat; a flea accelerates with a force of 135 *g* and is immediately invisible to the human eye.

Birds combine this acute vision with long-term memory and the ability to generalize. This would seem to make them invincible predators. But, paradoxically, these very abilities are the selective agents which have shaped and directed the evolution of insect defence policies.

The pressure is not all in one direction, though. Each advance by the insects is itself a selection pressure on birds to improve their vision and adroitness. Predator and prey are forever locked in an escalating arms race.

Mechanical defence: adult insects may protect themselves by mechanical devices, tough spines, and threatening postures.

199. (*Above*) A click beetle, *Chalcolepidius porcatus* (Fam.: Elateridae), Trinidad, on its back, showing the tension-loaded device by which it rights itself. The mechanism involves a flexible articulation between pro- and mesothorax. The black peg projecting backwards under the prothorax impinges on the forked tip of the mesothorax with a friction hold. Tension in two large muscles is released, the friction hold breaks, and the beetle leaps into the air with a loud click, startling a would-be predator.

200. (*Facing, top*) This nightmare in green, a bush locust, *Phymateus purpurescens* (Fam.: Pyrgomorphidae), Kenya, has a heavily-spined thorax, making it difficult to handle. But spines are not the first line of defence. If threatened, the bush locust suddenly spreads its brightly coloured hind wings as a warning of nastier things to come. If an inexperienced predator persists, the insect provides an unforgettable experience: it exudes a frothy, foul smelling, and distasteful fluid from between the thoracic plates.

201. (*Facing, bottom*) A threatened praying mantis, *Parasphendale agrionina* (Fam.: Mantidae), Kenya, rears up in an aggressive posture. The lightning reflexes, multiplicity of spines, and the gin-trap claws are a formidable armoury. Normally used for catching prey (see 33), the claws are equally effective in preventing the mantid from becoming prey itself; they can easily draw blood from a human if mishandled.

Defensive strategies in insect larvae: portable shelters, spiny barbs, and hairiness

202. A bagworm moth caterpillar, *Oiketicus* sp. (Fam.: Psychidae), feeds on a leaf in rain forest, Peru. The larva lives within a bag of leaf and twig fragments woven together with silk. The portable shelter is temporarily suspended from a twig by a girdle of silk. When it has eaten all the food within reach, the larva bites away the silk supporting the shelter and moves on, carrying its home, to a new source of food. The head and three succeeding segments are strikingly coloured; if disturbed, the larva darts in and out of its bag in a startling flash display.

203. These gregariously feeding leaf beetle larvae and pupae (Fam.: Chrysomelidae), Mexico, are armed with multi-spined, waxy outgrowths from the cuticle. The spines are secreted by cuticular wax glands and make the larvae unpleasant to handle for predators. The pupae retain part of the skin of the last larval instar, so as to remain protected by the spines.

204. (*Facing, top*) The hideous spiky spines of this 'baron and count' butterfly caterpillar, *Euthalia* sp. (Fam.: Nymphalidae), Malaysia, serve a dual purpose. First, the interlocking crown of thorns makes it virtually impossible for parasitic wasps, such as braconids or ichneumons, to get close enough to inject their eggs. Secondly, the orange tips of the main spines signal a warning to birds and lizards that they contain an irritant poison. The main spines are hollow and connect with cuticular poison glands. The spines break off readily when handled, penetrating a predator's skin and releasing the poison. This causes pain and/or a foul taste, swelling, and, in humans, a blistering rash.

205. (*Facing, bottom*) A beak full of fine, dense, and irritating hairs awaits any bird foolish enough to tackle this moth caterpillar (Fam.: Megalopygidae) from Mexico, though some birds, e.g. cuckoos can cope with hairy caterpillars.

Defence by ants and chemicals: some insects make offers which ants cannot refuse; others use chemical weapons

206. (*Facing*) A group of worker red ants, *Myrmica rubra* (Fam.: Formicidae), 'milk' honeydew from feeding aphids, *Aphis* sp., England. The ants stimulate the flow of honeydew by stroking the aphids with their antennae. The sweet honeydew is highly prized by the ants. In return, they protect the aphids against hoverfly larvae and ladybirds. But a small parasitic wasp, *Aphidius* sp. (Fam.: Braconidae) apparently goes unnoticed and lays an egg in an aphid.

207. The caterpillars of many of the 'blue' butterflies (Fam.: Lycaenidae) have evolved very close relationships with ants. As with the aphids, a sugary liquid is offered in return for protection by the ants. Here, an Australian lycaenid larva is attended by workers of the Green Weaver Ant *Oecophylla smaragdina* (Fam.: Formicidae). The fluid they crave is secreted by a gland in the caterpillar's 10th body segment. An ant solicits a droplet by stroking the glandular area with its antennae; the caterpillar responds by extruding the orifice of the gland and a droplet appears.

208. Chemical defence, in the form of an acrid, frothy secretion, is the forte of this Bornean tiger moth, *Rhodogastria* sp. (Fam.: Arctiidae). The moth gives an initial warning by suddenly spreading its wings to expose its brightly coloured abdomen, an invitation to leave the vicinity. If this fails, it exudes the foul yellow secretion from its thorax, accompanied by a hissing noise — an unforgettable experience.

I taste awful: very many insects advertise their distasteful nature with warning or aposematic colours

209. The harlequin patterns of this grasshopper duo, *Chromacris colorata* (Fam.: Acrididae), Mexico, signal loud and clear that these gregarious grazers are highly poisonous. Bright scarlet hindwings emphasize the point when they are exposed rapidly, a flash warning that birds and lizards ignore at their peril.

210. The bright livery of this desert grasshopper *Dactylotum* sp. (Fam.: Acrididae), Mexico, has the same message for would-be predators. Very often, the poisonous properties of an insect are derived from a toxic foodplant; the insect is able to tolerate poisons which the plant has evolved as an anti-herbivore device. Such insects are often ultra-specialists and feed on only one species of plant.

211. A Malaysian assassin bug, *Eulyes amaena* (Fam.: Reduviidae), runs over a tree trunk in search of insect prey. Its warning colours indicate not only its foul smell and taste, the products of thoracic scent glands, but also its awesome bite. The bug jabs its rostrum (see 53) into the flesh of an attacker, injecting saliva and causing excruciating pain.

212. (*Facing*) Warningly-coloured or aposematic insects are often gregarious: a large number of individuals enhances the overall visual effect on predators. Moreover, there is safety in numbers; the larger the assemblage, the smaller the risk to the individual. Here, a group of Seven-Spot Ladybirds, *Coccinella septempunctata* (Fam.: Coccinellidae), England, hibernate gregariously on a twig. This species usually hibernates in exposed places, often in hundreds of thousands, a vivid testimony to the efficiency of the aposematic colours. If handled roughly or attacked by a predator, they exude some bright orange haemolymph (blood) from their leg joints, a process called reflex bleeding. The liquid is sticky and very bitter to taste. It contains a substance called coccinelline, an alkaloid manufactured by the beetles themselves. The larvae of some species have a similar defence mechanism and exhibit reflex bleeding. Although coccinelline is effective against ants and many birds, there are reports that bears occasionally feed on ladybirds hibernating in snow fields. Perhaps in severe cold, the ladybirds cannot mobilize their defences.

We taste awful too: more toxic insects with warning or aposematic colours

213. Mating insects are particularly vulnerable to predators and often hide in inaccessible places. But these warningly-coloured frog-hoppers, *Cercopis vulnerata* (Fam.: Cercopidae), England, have no need to hide; they mate in the open, relying on their bright red and black liveries to advertise their nasty taste.

214. Black and red patterns are widespread among distasteful insects. Here, the pattern signals that this oil beetle, *Mylabris pustulata* (Fam.: Meloidae), Sri Lanka, is deadly poisonous. Its elytra (wing cases) contain cantharidin, a poison which causes violent itching and blistering of the skin. This is the basis of the aphrodisiac 'Spanish Fly' and is extracted from meloids by the pharmaceutical industry, though for different purposes.

215. A Magpie moth, *Abraxis glossulariata* (Fam.: Geometridae), England. Many day-flying moths are noxious and are warningly coloured. Birds soon learn to associate the colour pattern with an unpleasant experience. The caterpillars of the Magpie Moth have similar markings and are also distasteful. In one experiment, captive, birds were offered a mixture of palatable caterpillars and those of the Magpie Moth; the latter were avoided.

216. (*Facing, top*) The jewel-like pattern of this shield bug, *Steganocerus* sp. (Fam.: Scutellaridae), from Kenya, may be attractive to the human eye, but to insectivorous birds and lizards it signals a foul taste and revolting smell. The nymphs are even more brightly coloured and live in dense aggregations. *Steganocerus* spp. are sometimes pests of millet and sorghum.

217. (*Facing, bottom*) This warningly coloured Soldier Butterfly, *Danaus eresimus* (Nymphalidae: Danaiinae), Mexico, contains a mixture of chemicals called cardiac glycosides, which the caterpillars sequester from their food plants, species of Milkweed (*Asclepias*). The glycosides are bitter-tasting and cause vomiting in predators. Milkweeds evolved the manufacture of these chemicals to deter herbivorous insects and mammals; the Soldier, however, is immune to the glycosides.

Warning coloration in immature insects: many larvae employ chemical defences, often derived from their foodplants

218. The unforgettable livery of these Cinnabar Moth caterpillars, *Callimorpha jacobaeae* (Fam.: Arctiidae), England, reminds would-be predators that they should be avoided. Like those of the Soldier Butterfly (see 217), Cinnabar caterpillars sequester highly poisonous compounds from their host plants, in this case Groundsel and Ragwort (*Senecio vulgaris* and *S. jacobaea*). The poisons are pyrrolizidine alkaloids, a protection against grazing animals.

219. A combination of warning colours and sharp spines protects this moth caterpillar, *Dirphia molippa* (Fam.: Saturniidae), Trinidad. The nature of the chemical protection, if any, is unknown for this species, though warning colours usually enable one to predict its presence (but see 235–239). The branched spines (scoli) may contain an irritant substance (see 204).

220. These 'slug caterpillars' of a moth belonging to the family Limacodidae, Trinidad, are protected by poisonous spines. On contact, they inflict an excruciatingly painful sting. The warning effect of the colours is enhanced by the gregarious feeding habits of the young caterpillars (see 212). The slug-like appearance derives from the indistinct segmentation of the body and the virtual absence of legs.

221. Gregarious feeding is, perhaps, part of the defence strategy adopted by these conspicuously spined and coloured moth caterpillars, *Rohaniella sufferti* (Fam.: Saturniidae), feeding on a species of *Ochna*, in the Sokoke Forest, Kenya. The moth has been abundant here for at least 15 years and, so far, no parasites have bred from this population. The caterpillars sometimes completely defoliate the foodplants.

222. (*Above*) The safety-in-numbers principle is adopted by warningly coloured, early-instar nymphs of many species of grasshoppers. Here, nymphs of a *Taphronota* sp. (Fam.: Acrididae) feed gregariously in Kenya. They surround a larger, later instar nymph which has the adult colours. The adults and later nymphs feed singly; their green colour matches their background (crypsis). The use of different protective strategies at different times in the life-cycle is common in insects.

223. A cluster of shieldbug nymphs, *Lyramorpha* sp. (Fam.: Tessaratomidae), feeds on a leaf in rain forest, Queensland, Australia. The beautiful colours remind potential predators of their unpalatability. They produce a foul stench from glands in the abdomen, the four openings of which are clearly visible here. Some species of this family can eject an evil-smelling fluid a distance of 15–30 cm.

I am a wasp or a bee: many insects mimic the appearance and gait of stinging insects

224. Bees and wasps usually advertise the sting in their tails with a combination of yellow and black markings. It takes only one experience for an insectivorous bird to associate the colour pattern with a sharp pain. The insects therefore exploit the acute visual memory of birds. Many palatable insects gain protection by assuming the livery of venomous species.

225. This Wasp Beetle, *Clytus arietis* (Fam.: Cerambycidae), England, is a superb mimic of social wasps. The resemblance involves not only colour, but also gait: the beetle walks over leaves and flowers in an agitated, wasp-like manner. The resemblance to venomous insects by palatable ones is called Batesian mimicry, after H. W. Bates, the Victorian naturalist whose studies of South American butterflies first revealed the phenomenon.

226. Batesian mimics of wasps are found in many insect groups. Here, a hoverfly, *Chrysotoxum cautum* (Fam.: Syrphidae), England, basks on a flower of wild rose. Protected by their warning colours, wasp mimics can roost in exposed places with impunity. A few wasp-mimicking hoverflies are known to be distasteful in their own right and are therefore Müllerian rather than Batesian mimics (see 235–237).

227. The markings of this hoverfly, *Volucella zonaria* (Fam.: Syrphidae), France, resemble those of some social wasps, including the hornet, *Vespa crabro* (Fam.: Vespidae). The females of *Volucella zonaria* lay their eggs in the nests of social wasps, where the larvae scavenge on waste material, though towards the end of the colony's life they may begin to feed on the wasp larvae and moribund workers.

228. A bumblebee mimic, *Volucella bombylans* (Fam.: Syrphidae), England. There are several colour forms, each of which mimics the colour pattern of particular species of bumblebee. This one mimics those such as *Bombus lapidarius*, which are black with a bright orange tail. As in *Volucella zonaria*, the larvae are nest scavengers, in this case, in the nests of bumblebees. The mimicry of bumblebees does not confer free entry to the nests of their hosts for egg-laying purposes: female *Volucella* often lay eggs in the nests of species which they do not resemble. Rather, it is a simple case of Batesian mimicry.

229. Another colour form of *Volucella bombylans*. This mimics black, yellow, and white marked bumblebees such as *Bombus lucorum* and *B. hortorum*. As with most mimics of stinging insects, the resemblance goes far beyond colour pattern. Species of *Volucella* mimic the characteristic flight pattern of their models. *V. bombylans* makes a loud buzzing noise and, if disturbed, it adopts the typical bumblebee threat posture, raising one of its middle legs, accompanied by a loud buzz.

230. A chafer beetle, *Trichius fasciatus* (Fam.: Scarabaeidae), feeds at a thistle flower in the Forest of Dean, England. This beetle looks and sounds like a bumblebee, with its bumbling, noisy flight. It is active at flowers in summer, when bumblebees are most numerous and when most insectivorous birds have learned to associate this colour pattern with a painful sting.

Mutillid wasp mimics: many species mimic the wasps with the most powerful of all insect stings

231. A female mutillid wasp, *Lobotilla leucospila* (Fam.: Mutillidae), Kenya. The females are always wingless and the larvae are solitary external parasites of the larvae or pupae of other insects, especially other Hymenoptera (see 197). Mutillids are often called 'velvet ants' because of their velvety pubescence and rapid, ant-like gait. Although some North American species are called 'cow killers' because of their large size and agonizing sting, they pose no threat to livestock. The venom of mutillids causes massive swellings and it is not surprising that they are some of the most strikingly aposematic of all insects.

232. This assassin bug, *Ectomocoris* sp. (Fam.: Reduviidae), runs about in the same area in Kenya as the mutillid wasp shown above. With its pair of white spots and its rapid gait, the bug was almost indistinguishable from the mutillid wasp. It is not surprising that insects have come to mimic mutillids because ground-feeding insectivorous birds soon learn to avoid the fearsome sting of the wasps and leave well alone any insect which resembles them.

233. (*Facing, top*) Some insects are extremely accurate mimics of mutillid wasps. This ground beetle, *Eccoptera cupricollis* (Fam.: Carabidae), is such a good one that it was possible to identify which species of mutillid is the model, without seeing a specimen of the wasp. The beetle is one of several which mimic the widespread African mutillid, *Dolichomutilla guineensis*. The majority of the 10 000 or so mutillid species are tropical and a wide range of insects mimic them in all regions.

234. (*Facing, bottom*) Many spiders mimic mutillid wasps. This jumping spider, *Orsima formica* (Fam.: Salticidae), has a combination of colours and spots which is characteristic of many Malaysian species of Mutillidae. Note that while the spider's head is on the right, it is the abdomen, on the left, which mimics the mutillid's head, complete with false antennae. This is a curious example of reverse head mimicry (see 244–246).

Five into one will go: Müllerian and Batesian mimicry in South American Butterflies.

235. An Isabella Tiger, *Euides isabella* (Nymphalidae: Heliconiinae), Trinidad. It and the four others shown here have remarkably similar wing patterns, although they represent five subfamilies in two unrelated families, *E. isabella* and the next two species are unpalatable: predators learn to associate their colour patterns with a foul taste and nausea. These butterflies are therefore Müllerian mimics: the advantage of several poisonous species resembling each other lies in the fact that it reduces the number of colour patterns which predators must learn and therefore minimizes the number of individual butterflies sacrificed in the process. This explanation was first proposed by the naturalist Fritz Müller, in 1878, who, like Bates (see 225), studied South American butterflies. Modern research confirms Müller's interpretation and many complex Müllerian assemblages are now known in insects. *E. isabella* contains cyanide sequestered during larval life from the foodplants, passion flowers (*Passiflora* spp.).

236. The Sweet Oil, *Mechanitis isthmia* (Nymphalidae: Ithomiinae), in the same forest habitat in Trinidad. The larvae of this species sequester poisons from species of *Solanum*, members of the potato family, which contains such poisonous plants as Deadly Nightshade. *M. isthmia* has a slow, lazy flight and rests in shady places. It visits flowers in sun-spots in the forest.

237. A Lady Tiger, *Lycorea ceres* (Nymphalidae: Danaiinae), Trinidad. The larvae of this species sequester cardiac glycosides from their foodplant, the Milkweed, *Asclepias curassavica*. The toxins persist in the adult, which has a lazy flight, like that of the previous two species, and lives in damp, shady forest. Cardiac glycosides are powerful heart poisons and are used in minute quantities in the manufacture of certain drugs.

238. A mating pair of *Phyciodes philyra* (Fam.: Nymphalidae), in Mexico. This and the following species are palatable butterflies. They enjoy protection from bird predators by closely resembling the highly toxic species in 235–237; they are, therefore, Batesian mimics. The caterpillars of *P. philyra* feed on members of the daisy family, Compositae. This species is more closely related to the Small Tortoiseshell of Eurasia and North America than to the species it mimics.

239. This Tiger Pierid, *Dismorphia amphione* (Fam.: Pieridae), Trinidad, is a denizen of secondary forest. It not only mimics the wing markings of the three distasteful species, but also mimics the flight pattern of the danaiine, *Lycorea ceres*. The power of natural selection, in the form of bird predation, is demonstrated by the fact that this species is more closely related to the Cabbage White of Europe and North America than to its models.

Two heads are better than one: eyespots and reverse-head mimicry to startle and confuse predators

240. A caterpillar of the Large Elephant Hawkmoth, *Deilephila elpenor* (Fam.: Sphingidae), England, rears up and exposes two pairs of conspicuous eyespots. The eyespots are normally concealed beneath folds of skin while the caterpillar feeds. When disturbed, the head is retracted, unwrinkling the skin on the two fleshy segments behind the head and the eyespots are dramatically exposed. This startles a bird attempting to peck.

241. The majority of moths fly at night, when, with the exception of bats, there are few predators about. They are more vulnerable during the day, when they roost. Moths are therefore often drab and inconspicuous, like this *Automeris* sp. (Fam.: Saturniidae), Peru. In many species, though, the forewings conceal a surprise . . .

242. When the moth is disturbed, perhaps by a bird uncertain of what it has found, it raises its forewings, suddenly exposing the brighter coloured hindwings, each marked with an eyespot. This is repeated rapidly several times. The sudden appearance and disappearance of the eyespots is startling and frightens the bird away. This is an example of 'flash coloration' (see 280–283).

243. The eyespots on the hindwings of this Australian ringlet butterfly, *Hypocysta metirius* (Fam.: Satyridae), have a different function. They are permanently exposed and serve not to startle predators but act as decoys or false targets, directing a bird's attention (and bill) to a non-vital area. Living butterflies often have beak marks or nicks in the wings, showing that birds do aim for eyespots.

244. A Kenyan butterfly, *Abisara rogersi* (Fam.: Riodinidae) with eyespots and false antennae on the hindwings, giving the impression that the head is at the rear end of the body. The false antennae appear to twitch because the hindwings move up and down; the real antennae remain still. Moreover, the butterfly performs an about turn on landing; when it flies off again, it goes in the opposite direction to that expected. Experiments with small lizards have shown the effectiveness of this strategy.

245. Birds tend to make their first strike at the head; false head mimicry diverts their attention to non-vital parts of the body. It is commonest in the butterfly family Lycaenidae (Blues, Coppers, Hairstreaks) but is found in other families too. Here, a Zebra Butterfly, *Colobura dirce* (Fam.: Nymphalidae), Trinidad, drinks at muddy ground. The converging black stripes of the hindwings lead the eye to an eyespot and false head.

246. False-head mimicry is not confined to adult insects. This aggregation of unidentified Malaysian moth caterpillars add false-head mimicry to their warning coloration. They thus combine a warning, the safety-in-numbers principle, and deception. With the exception of the two individuals at the lower right, all the caterpillars are aligned with their heads to the top of the photograph. When disturbed, they rear up and wave their false heads *en masse*, presenting an intimidating spectacle.

We are not really here: we are sticks. Many insects avoid predators by resembling inedible objects

247. This bush cricket or katydid, *Tegra novaehollandiae* (Fam.: Tettigoniidae), Malaysia, lies doggo on a dead twig on the forest floor. With its mottled markings, it resembles a dead, lichen-covered twig. This is enhanced by the antennae and front pair of legs, which are held out in front, closely applied to the stick on which the insect rests; the hind legs are tucked beneath the wings.

248. One order of insects, the Phasmida, consists almost entirely of stick and grass stem mimics. Here, *Acanthoclonia paradoxa* rests on a forest fern in Trinidad. This species is so stick-like that it can sit safely on leaves in full view, like a fallen twig. It rarely, if ever, sits on lichen-covered branches, as so often falsely posed in picture books.

249. The inchworms or looper caterpillars of many geometer moths (Fam.: Geometridae) resemble twigs. They rest by day with their bodies at right angles to a branch or leaf petiole and are almost undetectable. In this Mexican species, the effect is curiously enhanced by the larvae of a parasitic braconid wasp, which resemble buds. They have finished feeding and have emerged to spin cocoons and pupate.

250. (*Facing, top*) A female, stick-mimicking praying mantis, *Oxyophthalmellus somallicus* (Fam.: Mantidae) (*right*), guards her cluster of nymphs (*left*), in Kenya. When poked with a stick, she always retreated away from her offspring, as if to distract attention from them. She returned to her guardpost by walking backwards, so that she faced the main stem of the bush, which would be the direction of attack by predators such as ants. Meanwhile, the nymphs remained motionless, resembling small thorns.

251. (*Facing, bottom*) An unidentified giant lacewing (Neuroptera: Ascalaphidae), Mexico, perches with its abdomen jutting out like a twig, behaviour also observed in an African species. This predator of aerial insects hunts at dusk and, like all the species on these and the next eight pages, rests during the day. If the resemblance to an inedible object is sufficiently close, then the insect can rest in exposed situations without attracting the attention of predatory birds.

252. A female treehopper or thorn bug (Homoptera: Membracidae) guards her eggs on an *Acacia* in the Tsavo National Park, Kenya. She resembles the *Acacia* thorns and thus is not detected by predatory birds. A female membracid usually hops away if disturbed but if she has a clutch of eggs to guard, she remains with them. Many species are attended by ants, rather like aphids. They exude a sweet, sugary fluid from the retractile anal tube. Although this female may be guarded by ants in return for honeydew, the ants do not respect her eggs and would eat them if they were left unattended. The ants shown here, though, are more interested in the nectar secreted by the *Acacia* from glands at the base of the thorns.

253. A cluster of treehoppers, *Antianthe expansa* (Homoptera: Membracidae), tended by ants in Trinidad. Like their aphid relatives, membracids have piercing and sucking mouthparts and tap the flowing sap of plants. Relatively large insects which feed in this way remain motionless for hours and may attract the attention of predators. The resemblance to green thorns, however, is protection enough for these bugs.

254. A seed-mimicking beetle, *Endustomus* sp. (Fam.: Tenebrionidae), Kenya, rests on the ground during the day. The resemblance to the keeled seeds of the tree family Combretaceae is so striking that the photographer collected them as such. The beetles lie motionless on the ground during the day among the seeds and are blown about with them by the wind. They are active only at night.

255. The elytra (wing cases) of this tortoise beetle, *Batanota bidens* (Chrysomelidae: Cassidinae), Trinidad, are fused and bear a long, central spine. Depending on where it rests, this beetle resembles a thorn or a seed.

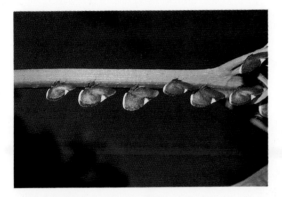

256. With their bodies all aligned in the same direction, these treehoppers, *Membracis tectigera* (Fam.: Membracidae), Trinidad, resemble a spike of withered flowers. The helmet-like pronotum is massively developed and protrudes forward to cover the head, and backwards to cover the abdomen. In other species, the pronotum resembles chewed leaf margins and many have spiny outgrowths with bulbous sections, resembling the bodies of ants.

257. A nymph of the flower mantid, *Pseudocreobotra ocellata* (Fam.: Hymenopodidae), Kenya. The outgrowths of the body and legs break up the mantid's outline and, together with the pink colour, make it resemble the petals of a flower. This not only makes it difficult for predators to see it but presumably makes it invisible to flower-visiting insects on which it feeds.

We are not really here: we are leaves. Many insects have evolved a protective resemblance to leaves

258. This grasshopper, *Plagiotryptus hippiscus* (Fam.: Eumastacidae), Kenya, relies on structure, colour, and posture to maximize its resemblance to a living green leaf. The thorax is laterally compressed to mimic the blade of a leaf. When the grasshopper is at rest, the raised hind legs and uptilted abdomen maintain the leaf-like profile.

259. Green leaves are a feature of nearly all parts of the Earth. It is not surprising, therefore, that many insects have independently evolved protective resemblance to leaves. Here, a bush cricket or katydid, *Phyllomimus* sp. (Fam.: Tettigoniidae), rests on leaves in a Malaysian rain forest. This superb mimic includes leaf veins and disease blotches in its repertoire of deception.

260. This bush cricket from Trinidad, *Pycnopalpa bicordata* (Fam.: Tettigoniidae), qualifies as the ultimate in leaf mimics. It not only resembles a leaf in shape but incorporates many features of a dead or dying leaf. It even mimics symptoms of fungal attack and blotch mines caused by a leaf-mining moth. Its resemblance to a leaf is so convincing that the insect is not confined to staying among leafy backgrounds.

261. A Walking Leaf, *Phyllium pulchrifolium* (Fam.: Phyllidae), New Guinea, is dorso-ventrally flattened into a leaf shape. This highly modified member of the stick insect order Phasmida also mimics an ageing leaf, with signs of decay and insect damage. *Phyllium* often suspends itself from a twig or leaf midrib by just two or three legs and rotates slowly, like a dead leaf moving in the wind.

262. This Malaysian grasshopper, *Chlorotypus* sp. (Fam.: Eumastacidae), is fully committed to resembling a dried, dead leaf. The populations of some species comprise green individuals which look like living leaves, as well as brown or grey, dead leaf mimics. Often, one or other colour form predominates according to season, a remarkable example of fine tuning to the relative proportions of available backgrounds.

263. A striking resemblance to dead leaves on the forest floor protects this pair of mating grasshoppers, *Ixalidium obscuripes* (Fam.: Acrididae), Kenya.

264. Many tropical cockroaches scavenge on the forest floor among dead leaves. Here, a species of *Rhabdoblatta* (Fam.: Blaberidae) scurries about among the dead leaves it resembles so well in the Makadara Forest in the Shimba Hills, Kenya.

265. Swaying slightly in the breeze, this female praying mantis, *Acanthops falcata* (Fam.: Hymenopodidae), Trinidad, is a striking mimic of a russet, crinkled dead leaf. So effective was the mimicry that the photographer initially overlooked it while observing other insects in the vicinity.

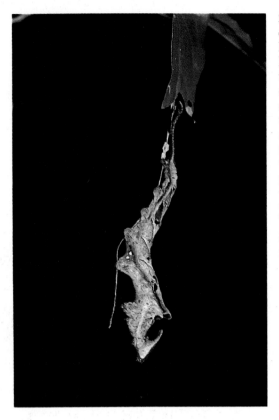

266. This crinkled leaf is really the caterpillar of the Silver King Shoemaker Butterfly, *Prepona antimache* (Fam.: Nymphalidae), Trinidad. It feeds during the night and is inactive by day. To enhance its protective resemblance, the caterpillar spins a silk thread to extend the midrib of a leaf. It then suspends itself on the thread, where it may twist and twirl in the breeze. This caterpillar is a very effective mimic of a particular pattern of leaf decay common among tropical trees, where death and withering proceed from the tip of the leaf to the base. Although widespread in tropical South America, the species of *Prepona* are rare butterflies.

267. (*Facing, top*) The Leaf Shoemaker Butterfly, *Anea itys* (Fam.: Nymphalidae), Trinidad, rests with its wings upright, so that the undersides are exposed. It presents a very thin, leaf-like profile. The wings are leafy in shape and resemble the colours of a yellowed, dead leaf, with mould blotching. The indented hind margin of the forewings gives the appearance of a leaf damaged by chewing insects. The deception is completed with a false midrib in the form of a dark stripe running down the middle of both wings. When the insect lands, it does so with a slow, fluttering flight, just like a dead leaf falling from a tree. It is therefore a superb example of how protective resemblance involves a combination of traits — colour, shape, posture, and gait. It also reflects the intensity of selection exerted by predatory birds.

268. (*Facing, bottom*) A Malaysian tussock moth, *Carniola* sp. (Fam.: Lymantriidae) takes dead leaf mimicry to the limit. Areas on both fore- and hindwings are devoid of scales so that the moth's own network of veins mimics that of a leaf skeletonized by decay. The great diversity of insects which mimic leaves both living and dead shows that natural selection can result in very different and unrelated insects evolving a similar response to a shared problem: staying alive despite the visual acuity of insectivorous birds.

We are not really here: bird droppings and stones may not be all that they seem

269. What could be more unappetizing to a bird than its own excreta? It is no surprise, therefore, that many insects and spiders have evolved a protective resemblance to bird droppings. This is worldwide in the young caterpillars of swallowtail butterflies, such as this King Page, *Papilio thoas* (Fam.: Papilionidae), in Trinidad. With its shape, marbled markings and shiny surface, the caterpillar is a superb mimic of a fresh bird dropping.

270. A nymph bush cricket or katydid (Fam.: Tettigoniidae), Trinidad, mimics a bird dropping. It has the same marbled coloration as the *Papilio* caterpillar, the white blotches mimicking the uric-acid paste typical of bird excreta. With hind legs splayed out and lying flat to the leaf, the insect resembles bird excreta which has been dropped from a great height and spread on impact. The shiny surface suggests a fresh dropping. Other mimics are matt in texture, so as to resemble dried excreta.

271. *(Facing, top)* Groundhoppers are always dull and drab in appearance and many are adapted for life in dry, stony places. This Bornean species of *Dolatettix* (Fam.: Tetrigidae), however, lives on stony ground and screes on the humid slopes of Mt Kinabulu. The thorax is hugely developed to form a domed helmet covering the body from head to abdomen. This gives the whole insect a rounded, pebble-like shape. This individual blended in with its background so well that it was only discovered when the photographer noticed that a 'stone' kept hopping away from him as he moved around the slope.

272. *(Facing, bottom)* Stony deserts harbour many insects and plants which have a protective resemblance to stones. This nymph of *Xanthippus corallipes* (Fam.: Acrididae), Arizona, has many of the features which have evolved independently in desert grasshoppers all over the world. The stone-grey colour matches the dominant background and the matt surface sculpture resembles that of a broken piece of weathered rock; the legs are tucked in and the antennae are held down flat against the face, so that the outline is broken up as little as possible. When disturbed, the hopper leaps away rapidly and becomes invisible again as it lands and re-assumes its motionless stance.

We are not really here: optical illusions in the struggle to remain unseen by predators

273. (*Above*) A caterpillar of the Sweet Potato Hawkmoth, *Agrius cingulatus* (Fam.: Sphingidae). Like many other caterpillars, it uses countershading to avoid detection. Light normally comes from above, so the lower surface of an animal appears darker than the upper surface, heightening its three dimensional nature, no matter how effective its camouflage. Many camouflaged animals counteract this effect by having darker pigmentation on their upper surfaces — hence countershading. It flattens their appearance and renders them invisible. However, hawkmoth larvae feed upside down, so it is the *undersides* which are paler and the *upper* surfaces which are darker. They therefore show reverse countershading, to which photographs scarcely do justice.

274. (*Facing, top*) Countershading plays a part, too, in the camouflage of this Kenyan grasshopper, *Humbe tenuicornis* (Fam.: Acrididae): the upper surface is generally darker than the lower, with a gradation in between. The mottled pattern blends well with the multicoloured sand grains of the arid wastes where the species lives. The grasshopper remains motionless for long periods, but, if disturbed, leaps away very rapidly, exposing brilliant yellow hindwings. This is an example of flash coloration, designed to startle a predator (see 280–283), and is often found in camouflaged insects as part of their package of defensive devices. Shrikes, however, are sometimes able to catch members of this species.

275. (*Facing, bottom*) This North American desert grasshopper, *Bootettix argentatus* (Fam.: Acrididae) feeds exclusively on the pungent smelling leaves of Creosote Bush, *Larrea tridentata*. Its dominant colour matches exactly the shade of green characteristic of *Larrea* leaves. The resemblance is enhanced by the curved, pale bands and blotches, which approximate to the highlights of the glossy leaves of the host plant. *Bootettix argentatus* is almost totally invisible so long as it remains in the highly dissected environment of *Larrea* leaves, but, out of its specialized context, it is a very conspicuous insect. This is true of all species which are camouflaged to match their background.

We are not really here; we are lichen-covered bark

276. The majority of cicadas (Fam.: Cicadidae) are cryptically coloured. This Malaysian species is mottled black, grey, and brown; it is well camouflaged against its background of lichen-covered bark. The mottling effect also breaks up the characteristic outline of the insect. This makes it difficult, if not impossible, for a bird to discern the limits of the cicada's body. Although rendered almost invisible, male cicadas are not entirely inconspicuous: it is unnerving to walk through a forest, deafened by the songs of cicadas, but unable to locate a single individual because of their superb camouflage and powers of ventriloquism.

277. This lasiocampid moth caterpillar rests by day on lichen-covered tree trunks. The mottled coloration blends impressively with that of the background. But just as important are the lateral flanges of the body, with their tufts of hairs splayed out, flat on the surface of the trunk. They break up the outline of the body very effectively. The photographer failed to see the caterpillar at first, although examining a spider just 3 cm away. When he did notice it, he initially thought it was a lichen-covered section of spider's web, until he touched it. And it moved!

278. (*Facing, top*) The breaking up of a characteristic outline is common to all lichen mimics. Here, a bush cricket or katydid, *Cymatomera denticollis* (Fam.: Tettigoniidae), roosts on a lichen-covered tree trunk in Kenya. The knobbly outgrowths of the legs resemble the encrusted bark. Stillness and posture are important, too, in surviving the dangers of resting by day; the antennae and forelegs are flattened against the trunk.

279. (*Facing, bottom*) A group of remarkably camouflaged shield bugs, *Coriplatus depressus* (Fam.: Pentatomidae), on a lichen-covered tree trunk in Peru. The assemblage includes eggs, nymphs, and adults. One of the latter is a yellow female, mating with a normal coloured male. Inappropriate colour varieties such as this are conspicuous and are soon attacked by birds, an example of natural selection in action.

Surprise, surprise! Flash displays of startling colour in otherwise cryptic insects

280. A female bush cricket (katydid), *Neobarrettia vannifera* (Fam.: Tettigoniidae) wanders over its foodplant in Tamaulipas State, Mexico. Note the long ovipositor or egg-laying tube. With its green colour and stealthy movements, this species is well-matched with its background. But . . .

281. . . . if disturbed by an inquisitive bird or hungry lizard, it opens its forewings to reveal a polka-dot pattern on the membranous hindwings. This flash display is repeated rapidly and startles and inhibits the predator.

282. (*Facing, top*) Two butterflies, *Vanessula milca* (Fam.: Nymphalidae) feed at urine-soaked ground in the Kakamega Forest, Kenya. They rest with their wings closed over their bodies, exposing the undersides of the wings, which resemble dead leaves on the forest floor. But . . .

283. (*Facing, bottom*) . . . at the first sign of danger, they flick open their wings to expose the strikingly coloured upper surfaces. This is repeated rapidly and is an effective flash display. Defence by surprise is very widespread and is not confined to insects. It is found in fish, amphibia, reptiles, birds, and mammals, often associated with drab or camouflaged animals. Thus, it is defence on the principle of the least expected, a last resort if the first line of defence, camouflage, is breached by a lucky or unusually acute predator.

A talent to deceive: predators mimicking prey . . . and a **mystery**

284. Cotton stainer bugs, *Dysdercus* sp. (Fam.: Pyrrhocoridae), mating and feeding on cassava in Kenya. Species of *Dysdercus* are widespread in warm, cotton-growing areas. They are called cotton stainers because the feeding holes they pierce in cotton bolls allow a fungus, *Nematospora* sp., to enter and discolour the cotton. The heraldic markings of the bugs are aposematic, signalling to predators that they are distasteful. But this is not effective against one specialized enemy . . .

285. The assassin bug, *Phonoctonus nigrofasciatus* (Fam.: Reduviidae), feeds on the cotton stainer and resembles it very closely. This is not a form of stealth, enabling it to attack its prey. Rather, it is a form of Mullerian mimicry (see 235–237): like the cotton stainer, *P. nigrofasciatus* has stink glands and it can also inflict a painful bite. Thus, by presenting the same appearance, birds have to learn fewer colour patterns and therefore make fewer damaging experiments. It thus pays both predatory bug *and* its prey to present a unified front. A similar relationship exists between the assassin bug *Harpactor costalis* and *Dysdercus* in India.

286. (*Facing, top*) In this jumping spider, *Cosmophasis* sp. (Fam.: Salticidae), Kenya, resemblance to its ant prey *is* very clearly an adaptation for stealth. By mimicking the structure of worker ants, in which perception by touch is important, it fools its prey into thinking it is one of them. This is an essential and highly successful ploy, because ants are quite capable of killing the spider.

287. (*Facing, bottom*) It is a mystery just why this remarkable weevil, *Talanthia phalangium* (Fam.: Curculionidae), Malaysia, has such stilt-like legs. It is very similar to a beetle from the Kalahari Desert, where long legs keep the body away from the scorching sand. But this cannot be the case in rain forest. The weevil's rapid gait resembles that of certain distasteful harvestmen (phalangids) and herein may lie the answer.

Further Reading

The following texts are arranged to follow the seven introductory essays of this book. The texts for essays 3–5 are combined because there is broad overlap between them. The extensive bibliographies in the listed texts will provide access to the primary research literature.

1 Success in diversity

Alcock, J. (1975). *Animal behaviour, an evolutionary approach*. Sinauer Associates, Sunderland, Mass.

Baerends, G., Beer, C., and Manning, A. (eds.) (1975). *Function and evolution in behaviour. Essays in honour of Prof. Niko Tinbergen.* Oxford University Press.

Chapman, R. F. (1982). *The insects: structure and function*, 3rd. edn. English Universities Press, London.

Corbet, P. S. (1962). *A biology of dragonflies.* Witherby, London.

—— Longfield, C., and Moore, N. W. (1960). *Dragonflies* (The New Naturalist Series). Collins, London.

Crowson, R. A. (1981). *The biology of the Coleoptera.* Academic Press, London.

Daly, H. V., Doyen, J. T., and Erlich, P. R. (1978). *Introduction to insect biology and diversity.* McGraw-Hill, New York.

Eisner, T. and Wilson, E. O. (eds.) (1977). *The insects. Readings from Scientific American.* W. H. Freeman, San Francisco, Calif.

Evans, M. E. G. (1975). *The life of beetles.* George Allen & Unwin, London.

Ford, E. B. (1957). *Butterflies*, 3rd ed. (The New Naturalist Series) Collins, London.

—— (1972). *Moths,* 3rd edn (The New Naturalist Series) Collins, London.

Johnson, C. G. (1969). *Migration and dispersal of insects by flight.* Methuen, London.

Krebs, J. R. and Davies, N. B. (1978). *Behavioural ecology: an evolutionary approach.* Blackwell Scientific, Oxford.

Matthews, R. W. and Matthews, J. R. (1978). *Insect behaviour.* Wiley Interscience, New York.

Mound, L. M. and Waloff, N. (eds.) (1978). *Diversity of insect faunas*, 9th Symposium of the Royal Entomological Society. Blackwell Scientific, Oxford.

Oldroyd, H. (1964). *The natural history of flies* (World Naturalist Series) Weidenfeld and Nicolson, London.

Owen, D. F. (1971). *Tropical butterflies.* Oxford University Press.

Rainey, R. C. (ed.) (1975). *Insect flight.* 7th Symposium of the Royal Entomological Society. Blackwell Scientific, Oxford.

Richards, O. W. and Davies, R. G. (Revisers) (1977). *Imms' general textbook of entomology*, 10th ed. Chapman and Hall, London.

Ross, H. H., Ross, C. A., and Ross, J. R. P. (1982). *A textbook of entomology.* 4th edn. Wiley, New York.

Smith, K. G. V. (ed.) (1973). *Insects and other arthropods of medical importance.* Trustees of the British Museum (Natural History), London.

Southwood, T. R. E. (ed.) (1968). *Insect abundance.* 4th Symposium of the Royal Entomological Society. Blackwell Scientific, Oxford.

Wigglesworth, V. B. (1964). *The life of insects* (World Naturalist Series). Weidenfeld and Nicolson, London.

—— (1972). *The principles of insect physiology*, 7th edn. Chapman Hall, London.

2 Food and feeding

Emden, H. F. van (ed.) (1973). *Insect–plant relationships.* 6th Symposium of the Royal Entomological Society. Blackwell Scientific, Oxford.

Gilbert, L. E., Raven, P. R. (eds.) (1975). *Co-evolution of animals and plants.* University of Texas Press, Austin, Tex.

3 Finding a mate
4 Investments for the future: egg-laying and parental care
5 The social option

Barash, D. 1979. *Sociobiology: the whisperings within.* Harper and Row, New York.

Bell, W. J. and Cardé, R. T. (eds.) (1984). *The chemical ecology of insects.* Chapman and Hall, London.

Birch, M. C. (ed.) (1974). *Pheromones.* Frontiers of Biology no. 32. North Holland, Amsterdam.

—— Haynes, K. F. (1982). *Insect pheromones.* Studies in Biology no. 147. Edward Arnold, London.

Breed, M. D., Michener, C. D., and Evans, H. E. (eds.) (1982). *The biology of social insects.* Proceedings of the 9th Congress of the International Union for the Study of Social Insects, Boulder, Colorado, 1982. Westview Press, Boulder, Colo.

Blum, M. S. and Blum, N. A. (eds.) (1979). *Sexual selection and reproductive competition in insects.* Academic Press, New York.

Butler, C. G. (1954). *The world of the honeybee* (New Naturalists Series). Collins, London.

Clutton-Brock, T. H. and Harvey, P. H. (eds.) (1978). *Readings in sociobiology.* W. H. Freeman, Reading.

Dawkins, R. (1976). *The selfish gene.* Oxford University Press.

—— (1982). *The extended phenotype. The gene as the unit of selection.* W. H. Freeman, Oxford.

Evans, H. E. 1966. *The comparative ethology and evolution of the sand wasps.* Harvard University Press, Cambridge, Mass.

—— West-Eberhard, M. J. (1973). *The wasps.* David and Charles, Newton Abbot.

Free, J. B. and Butler, C. G. (1957). *Bumblebees* (New Naturalist Series). Collins, London.

Frisch, K. von (1967). *The dance language and orientation of bees.* (Trans. from the German by L. E. Chadwick). Belknap Press of Harvard University Press, Cambridge, Mass.

Heinrich, B. (1979). *Bumblebee economics.* Harvard University Press, Cambridge, Mass.

Hermann, H. R. (ed.). (1979–1982). *Social insects,* Vols. I–IV. Academic Press, New York.

Maynard-Smith, J. (1978). *The evolution of sex.* Cambridge University Press.

Michener, C. D. (1974). *The social behaviour of the bees: a comparative approach.* Belknap Press of Harvard University Press, Cambridge, Mass.

Oster, G. F. and Wilson, E. O. (1978). *Caste and ecology in the social insects.* Monographs in Population Biology 12. Princeton University Press, NJ.

O'Toole, C. and Raw, A. (in press). *A murmur of bees.* Oxford University Press.

Spradbery, J. P. (1973). *Wasps. An account of the biology and natural history of solitary and social wasps.* Sidgwick & Jackson, London.

Thornhill, R. and Alcock, J. (1983). *The evolution of insect mating systems.* Harvard University Press.

Williams, G. C. (1975). *Sex and evolution.* Princeton University Press, NJ.

Wilson, E. O. (1971). *The insect societies.* Belknap Press of Harvard University Press, Cambridge, Mass.

—— (1975). *Sociobiology, the new synthesis.* Belknap Press of Harvard University Press, Cambridge, Mass.

6 Parasites, nest scavengers, and cuckoos

Askew, R. R. (1971). *Parasitic insects.* Heinemann Educational, London.

Clausen, C. P. (1940). *Entomophagous insects.* McGraw-Hill, New York.

Varley, G. C., Gradwell, G. R., and Hassell, M. P. (1973). *Insect population ecology: an analytical approach.* Blackwell Scientific, Oxford.

7 Defence in a hostile world

Cott, H. B. (1940). *Adaptive coloration in animals.* Methuen, London.

Dobzhansky, T. (1970). *Genetics and the evolutionary process.* Columbia University Press.

Edmunds, M. (1974). *Defence in animals.* Longmans, London.

Kettlewell, B. (1973). *The evolution of melanism; the study of a recurring necessity.* Clarendon Press, Oxford.

Owen, D. F. (1980). *Camouflage and mimicry.* Oxford University Press.

Sheppard, P. M. (1967). *Natural selection and heredity.* Hutchinson, London.

Wickler, W. (1968). *Mimicry in animals and plants.* (Trans. from the German by R. D. Martin.) World University Library, Weidenfeld and Nicolson, London.

Index

595.7051 O'Toole,
OTO   Christopher.

Insects in camera

DATE			